RECENT RESEARCHES ON THERMOLYSIS OF ENERGETIC MATERIALS

RECENT RESEARCHES ON THERMOLYSIS OF ENERGETIC MATERIALS

GURDIP SINGH, CHANDRA PRAKASH SINGH
AND PREM FELIX SIRIL

Nova Science Publishers, Inc.
New York

For permission to use material from this book please contact us:
Telephone 631-231-7269; Fax 631-231-8175
Web Site: http://www.novapublishers.com

LIBRARY OF CONGRESS CATALOGING-IN-PUBLICATION DATA

Gurdip Singh.
Recent researches on thermolysis of energetic materials / Gurdip Singh, Chandra Prakash Singh, Prem Felix Siril, authors.
p. cm.
ISBN 978-1-60456-222-4 (hardcover)
1. Salts--Thermal properties. I. Singh, Chandra Prakash. II. Siril, Prem Felix. III. Title.
TP230.G87 2008
662'.2--dc22

2007052122

Published by Nova Science Publishers, Inc. ✦ New York

CONTENTS

Preface		**vii**
Chapter 1	High Energetic Materials	**1**
Chapter 2	Thermolysis of Ammonium Perchlorate	**23**
Chapter 3	Chemistry and Thermolysis of Salts of Perchloric Acid	**53**
Chapter 4	Chemistry and Thermolysis of Salts of Nitric Acid	**61**
Chapter 5	Chemistry and Thermolysis of Polynitro Compounds	**71**
Chapter 6	Thermolysis of Transition Metal Nitrate Complexes	**89**
Chapter 7	Thermolysis of 5-Nitro-2,4-Dihydro-3H-1,2,4-Triazole-3-One (NTO) and Its Salts	**119**
Chapter 8	Thermolysis of Plastic Bonded Explosives	**153**
Index		**171**

PREFACE

Thermolysis is one of the most widely studied aspects of energetic materials. This is because any accidental hazardous stimulii usually triggers off a thermal event and thus the thermal stability is a critical safety aspect for energetic materials. Moreover, the chemical products of thermolysis influence the performance of energetic materials. Thus, studies on thermolysis gives valuable inputs about the safety and performance, the two most crucial parameters that govern the applcation of energetic materials. There are a number of review articles dealing with the thermochemistry of individual energetic compounds. However, there is no single source of literature for a new entrant in the research field of thermolysis of energetic materials to start learning about the advances made in this research field. Hence we have attempted to cover the research done on this well advanced field of thermolysis of a wide variety of energetic materials. The book does not cover exhaustively all the work done on thermolysis of all energetic materials, but instead, we discuss thermolysis of a number of representative energetic materials from different classes. We made sincere effort towards maintaining the language and the content as simple as possible so that it can be used as a material for teaching students. However we hope that the scientific content is even interesting to the experts who are working in this and related fields.

The first chapter gives an overview about various classes of energetic materials and various thermoanalytical techniques that are usually employed for the study of thermolysis. Other chapters deals with the chemistry and thermolysis of AP, AN, salts of nitric acid, salts of perchloric acid, polynitro compounds, NTO and salts of NTO. Thermolysis of plastic bonded explosives is discussed in the last chapter.

Chapter 1

HIGH ENERGETIC MATERIALS

ABSTRACT

An introduction to high energetic materials and their types are given. The properties necessary for the safe and reliable handling of explosives have been discussed. Some conventional propellants, explosive and pyrotechnics compositions are also given. Some of the most common thermoanalytical techniques that are used for analysing energetic materials are also described in this chapter.

1.1. INTRODUCTION

Energetic materials are the nucleus of modern warfare and one of the most intriguing class of compounds which can act as a source of high chemical energy. An energetic material [1,2] is a chemical substance or mixture of substances which when raised to a sufficient high temperature whether by direct heating, friction, impact, shock, spark or flame, suddenly undergoes a very rapid chemical transformations with evolution of large amount of heat and gases, thereby exerting high pressure on the surroundings. In other words, it may be defined as a material that can undergo very rapid self propagation decomposition to release its potential energy and produces a sudden outburst of heat and volume of gases considerably greater than the original volume of the energetic material. Energetic material may be solid, liquid or gas and the products of explosion / ignition are gases or mixture of gases and solid. HEMs have two parts in their molecules:

(I) Explosophores or Plosophoric Group

The group which themselves produce explosive properties of any substance is called explosophores or plosophoric group [3]. The following groups are the explosophores-

i. $–NO_2$ and $-ONO_2$ in both organic and inorganic substances.
ii. –N=N and –N=N=N- in inorganic and organic azides.
iii. $–NX_2$ (X = halogen).
iv. –N=C in fulminates.

v. $-OClO_2$ and $-OClO_3$ in inorganic and organic chlorates and perchlorates respectively.
vi. –O-O- and –O-O-O- in inorganic and organic peroxides and ozonides respectively.
vii. (vii)-C≡C- in acetylene and metal acetylides.
viii. M-C (metal bonded with carbon) in some organometallic compounds.

Plosophores are divided into two classes i.e. primary and secondary. Primary plosophores include nitrate esters, aromatic and aliphatic nitro groups and the nitramine groups. The secondary plosophores comprise groups such as azo, azide, nitroso, peroxide, ozonide and perchlorate etc.

(II) Auxoplosive Groups

The groups which do not themselves produce explosive properties but may influence them are called auxoplosive groups. The hydroxyl, carbonyl, chlorine, sulphur, ether, oxygen and amine etc. are some examples of such groups.

1.2. Classification of High Energetic Materials

High energetic materials can broadly be classified into three classes- explosives, propellants and pyrotechnics on the basis of their application.

1.2.1. Explosives

Explosives are regarded as a means of storing energy, which can be liberated on appropriate stimulus to drive the weapons. In other words, any number of unstable chemicals and mixture of compounds, which may be capable of exploding with a range of degrees of violence, is called as explosives. Gunpowder was the first explosive recorded in a Chinese manuscript around 1040 AD and was described under the generic name of '*huo yao*'.

1.2.1.1. Classification of Explosives

Eplosives can be classified on the basis of several ways:

1.2.1.1.1. Classification on the Basis of Different Groups / Chemical Nature

Explosives can be classified (Scheme 1) both on the basis of their chemical nature and uses. From chemical point of view, these are individual substances or mixtures.

The individual substances are further divided into

- Nitro compounds
- Nitrate Esters
- Nitramines
- Chlorates and perchlorates
- Azides

- Various compounds capable of producing an explosion (fulminates, acetylides, tetrazene, peroxides etc.)

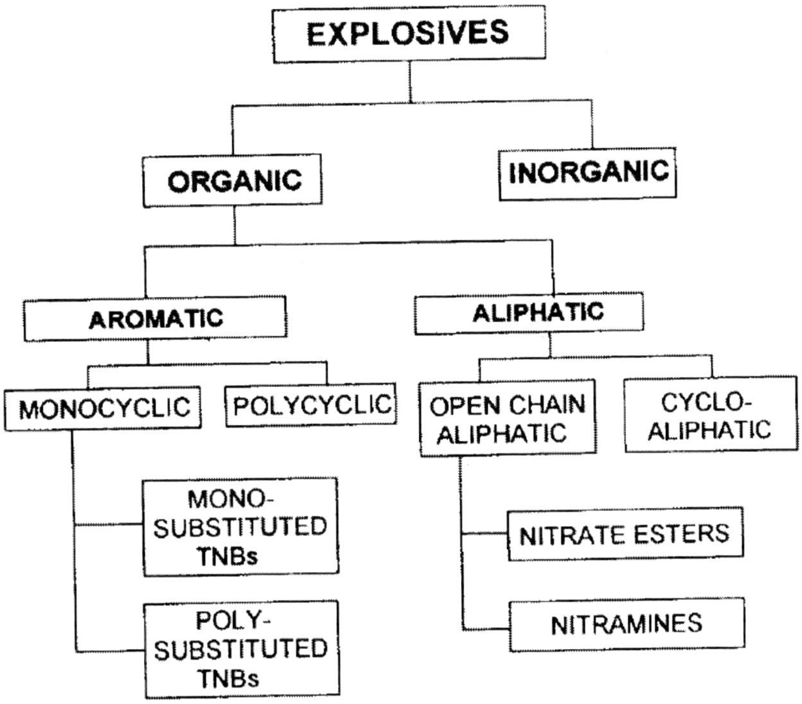

Scheme 1. Classification of Explosives on the basis of different groups / chemical nature.

1.2.1.1.2. Low and High Explosives

A low explosive is a combustible substance that decomposes rapidly (deflagrate) but does not explode under normal conditions. However, under certain conditions it is possible for them to detonate, usually when combined with high explosives. Low explosives are generally employed as propellants. They undergo deflagration at rates that vary from a few centimeter per second to ~400 meters per second. Examples are smokeless powders and pyrotechnics such as flares and illumination devices.

High explosives undergo detonation at rates of 1000 to 9000 meter per second. Such explosives are normally employed in mining, demolition and military warheads. Most of the low explosives are mixtures whereas most of the high explosives are compounds, but with some exceptions.

1.2.1.1.3. Primary, Secondary and Tertiary Explosives

Primary explosives are extremely sensitive to shock, friction and heat, they will burn rapidly or detonate if ignited.

Secondary explosives or base explosives are relatively insensitive to shock, friction and heat. These are sometimes added in small amounts to blasting caps to boost their power. Dynamite, TNT, RDX, PETN, HMX etc. are secondary explosives. PETN is often considered as benchmark compound in classification of explosives. The materials which are more sensitive than PETN being classified as primary explosives.

Tertiary explosives, also called blasting agents are so sensitive to shock that they can not be reliably detonated by practical quantities of primary explosives, and instead require an intermediate explosive booster of secondary explosive. Examples are ammonium nitrate / fuel oil mixture (ANFO) and slurry or 'wet bag' explosives. These are primarily used in large scale mining and construction operations.

1.2.1.1.4. Plastic Bonded Explosives (*PBXs*)

The general approach to maximizing the energy of the explosive while minimizing the energy of the sensitivity lies in the use of RDX or HMX or any of the newer high performance explosive after thoroughly coating them with a plasticised polymer. The function of the binder is to protect the explosive crystals from quick initiation due to outer stimuli. This is realized by coating the entire crystal surfaces of the explosive by a thin polymer layer. This layer can absorb the accidental energy pulses and distribute them evenly over the volume of the PBX, thereby preventing local decomposition and initiation. Another important function of the binder is its high mechanical strength: the binder will create a void free matrix between the explosive particles. The so formed PBX has a high mechanical strength that prevents crack formation even under high pressure conditions.

The polymer matrix gives strength to the product. A flexible and elastic polymer can reduce shock sensitivity by spreading local impulse over a large area and thereby reduce the probability of initiation. More importantly it reduces fatal detonation. Conventional as well as energetic polymers are being used as binders for PBXs. Some of the common plastic binders are presented in table 1.

Table 1. Binders used in plastic bonded explosives

Material	Description
Estane 5702-F1	Polyurethane solution system
Viton A	Vinylidene fluoride/hexafluoropropylene copolymer,60/40 wt%
BDNPA-F	Bis(2,2-dinitropropyl)acetal/bis(2,2-dinitropropyl)formal, 50/50 wt%
CEF	Tris-β-chlorethyphosphate
Kel-F 800	Chlorotrifluoroethylene/vinylidine fluoride copolymer, 3:1
FPC 461	Vinyl chloride/chlorotrifluoroethylene copolymer, 1.5:1
Polystyrene	Styrene polymer, 100%
FEPO	Bis(2-fluoro-2,2-dinitroethyl)formal

The binder is usually precipitated out of solution in the preparation process such that it coats the explosive crystals. Agglomerates of these coated crystals form pressing 'beads'. The beads are then either die pressed or isostatically pressed at high temperatures as high as 120^0C. Pressures from 1 to 20 kPsi then produce billets or pellets, with density as high as 97% of the theoretical maximum density (TMD). Table 2 lists various PBXs in common use in the U.S. Department of Energy's weapons laboratories.

Table 2. List Various PBXs

Name	Explosive ingredient	Binder Ingredient	Colour
LX-14-0	HMX, 95.5%	Estane 5702-F1, 4.5%	Violet spots on white
LX-10-0	HMX, 95%	Viton A, 5%	Blue-green spots on white
LX-10-1	HMX, 95.5%	Viton A, 5.5%	Blue-green spots on white
PBX-9501	HMX, 95%	Estane 2.5%; BDNPA-F, 2.5%	White
PBX-9404	HMX, 94%	NC, 3%; CEF, 3%	White or blue
LX-09-1	HMX, 93.3%	BDNPA, 4.4%; FEPO, 2.3%	Purple
LX-09-0	HMX, 93%	BDNPA, 4.6%; FEPO, 2.4%	Purple
LX-07-2	HMX, 90%	Viton A, 10%	Orange
PBX-9011	HMX, 90%	Estane and 5703-F1, 10%	Off-white
LX-04-1	HMX, 85%	Viton A, 15%	Yellow
LX-11-0	HMX, 80%	Viton A, 20%	White
LX-15	HNS-1, 95%	Kel-F 800, 5%	Beige
LX-16	PETN, 96%	FPC 461, 4%	White
PBX-9604	RDX, 96%	Kel-F 800, 4%	
PBX-9407	RDX, 94%	FPC 461, 6%	Black or white
PBX-9205	RDX, 92%	Polystyrene, 6%, DOP, 2%	White
PBX-9007	RDX, 90%	Polystyrene, 9.1%; DOP, 0.5%; Rosin, 0.4%	White or mettled gray
PBX-9010	RDX, 90%	Kel-F 3700, 10%	White
PBX-9502	TATB, 95%	Kel-F 800, 5%	Yellow
LX-17-0	TATB, 92.5%	Kel-F 800, 7.5%	Yellow
PBX-9503	TATB, 80%; HMX, 15%	Kel-F 800, 5%	Purple

Barring their extensive applications in munitions, PBXs also find applications in flight of launch vehicles. Pyrovalves, cable cutters, explosive latches, explosive zip, flexible linear shaped cords etc. are used for several functions. Pyrovalves are used to connect high-pressure line to downstream element otherwise positively disconnected because of safety requirements. Explosive bolts are used to separate the elements on demand. Flexible linear cords are used for stage separation and also vehicle destruct systems which need to get activated in case the vehicle deviates from the intended path.

1.2.1.2. Properties of Explosives

For an explosive to be of practical value it must be adequately safe and reliable and should possess the following properties:

- Stability under anticipated storage conditions i.e. should be unaffected by the environment or other materials stored with it
- Burn, explode or detonate only when required in use
- Should be sufficiently sensitive to be initiated as required and precisely when required
- The initiation stimulus should be small compared with the output of the explosive
- Must be capable of doing useful work on the target.

The effectiveness of an explosive depends on two factors. The first is the amount of energy available in the explosive and secondly the rate of release of the available energy when the explosion occurs. To measure the effectiveness of different explosives, a variety of performance parameters may be used such as:

- Heat of explosion
- Temperature of explosion
- Pressure of explosion
- Power index
- Rate of burning
- Detonation velocity (VOD)
- Detonation pressure

A key feature of performance of explosives is the dependence of detonation pressure on density (the density terms appears in the expression raised to the second power). The detonation velocity is inversely dependant on molecular weight of decomposition products. Thus, the best performing explosives will have large heats of decomposition, high density and decomposition products with the lowest possible molecular weight.

1.2.1.3. Safety Aspects of Explosives

Safety of explosives is of utmost importance and is usually associated with sensitivity of HEM to various extraneous stimuli such as shock, impact, friction, spark etc. Thermal stability is intimately related to sensitivity of explosives, as the above stated hazardous stimuli triggers off a thermal event, which is the main cause of initiation of any explosive [4]. Improved thermal stability generally simplifies production procedures, increases shelf-life of munitions and weapons and decreases their vulnerability to accidental initiation. Additionally, there are typical service applications (e.g. munitions carried under the wings of supersonic aircraft) wherein explosives are subjected to higher temperatures. For these reasons, acceptable thermal stability for the applications envisaged has long been a guiding factor for the chemist synthesizing new energetic materials and the formulators of new energetic compositions. There are four general chemical approaches to confer thermal stability to explosive molecules:

i. Introduction of an amino group
ii. Condensation with a triazole ring
iii. Salt formation and
iv. Introduction of conjugation.

A common formulation strategy used to develop insensitive munition explosive is to replace some of the conventional ingredients such as RDX or HMX with less sensitive ingredients such as NTO and nitro guanidine.

1.2.1.4. Some Conventional Explosive Compositions

Common explosives are composed of mixtures of an oxidizer and a fuel. Some common compositions are given in table 3.

Table 3. List of some Conventional Explosive Compositions

Name	Description
Black powder	Potassium nitrate, charcoal and sulfur
Flash powder	Fine metal powder (usually aluminium or magnesium) and a strong oxidizer (e.g. potassium chlorate or perchlorate)
Ammonal	Ammonium nitrate and aluminium powder
Armstrong's mixture	Potassium chlorate and red phosphorus. This is a very sensitive mixture. It is a primary high explosive in which sulfur is substitute for some or all phosphorus to slightly decrease sensitivity.
Sprengel explosives	A very general class incorporating any strong oxidizer and highly reactive fuel.This name is generally used for the mixtures of chlorates and nitroaromatics. Some others are: (i) *ANFO:* Ammonium nitrate and fuel oil (ii) *Cheddites:* Chlorates or perchlorates and oil *(iii) Oxyliquits:* Mixtures of organic materials and liquid oxygen
Nitroglycerin	A highly unstable and sensitive liquid, known as dynamite when mixed into sawdust, powdered silica or most commonly diatomaceous earth, which act as stabilizers.
Acetone peroxide	A very unstable white organic peroxide
TNT	Yellow insensitive crystals that can be melted and cast without detonation
Nitrocellulose	A nitrated polymer which can be a high or low explosive depending on nitration level and conditions.
RDX, PETN, HMX	Very powerful explosives which can be used pure or in plastic explosives

The compositions given in table 3 covers majority of the explosive materials. But a practical explosive will often contain small percentage of other materials. Plastics and polymers may be added to form plastic bonded explosives (PBXs). Waxes may be incorporated to make them safer to handle. Aluminium powder may be introduced to increase total energy and blast effect. Explosive compounds are also often alloyed. HMX or RDX powders may be mixed with TNT to form Octol or Cyclotol.

1.2.2. Propellants

Propellants are most important chemical formulations which are power behind rockets, missiles and launch vehicles. The energetic materials which after combustion produces hot gaseous products ejected from the nozzle to produce forward thrust to propulsion units is called propellant. Broadly propellants can be classified into four classes:

- Solid propellants
- Liquid propellants
- Hybrid propellants
- Gelled propellants

1.2.2.1. Solid Propellants

Physical state of these propellants is solid. The list of ingredients for solid propellants are presented in table 4.

Table 4. List of Solid Propellant Ingredients

Category of ingredients	Name/abbreviation of ingredients
DOUBLE BASE PROPELLANTS	
Energetic Plasticizers	NG, TMETN (trimethylol ethane trinitrate), TEGDN (triethylene glycol dinitrate)
Non-energetic Plasticizers	DEP (Diethyl phthalate), TA (triacetin), other organic phthalates (DBP, DOP etc)
Binders	Nc (nitrocellulose)
Stabilisers	EC or Carbamite (Symdiethyldiphenyl urea), 2-NDPA (2-nitro diphenyl amine)
Burning rate catalysts	Lead salts (lead stearate, lead salicylate, lead citrate), copper salts (copper stearate and salicylate)
High energy additives	RDX, HMX, NQ (nitroguanidine)
Opacifiers	Carbon black
Flame suppressants	KNO_3, K_2SO_4
Mettalic fuels	Al
Combustion instability suppressants	Al, Zr, ZrC (Zirconium carbide)
COMPOSITE SOLID PROPELLANTS	
Oxidisers	AP, AN, NP, KP, RDX, HMX
Binders	PU (Polyurethane), PBAN, CTPB, HTPB
Plasticisers	DOA (Dioctyl adipate), IDP (Idodecyl pelargonate), DOP
Metallic fuels	Al, Mg, B, Zr, Ti
Burn rate catalysts	Fe_2O_3, n-butyl ferrocene, LiF, Oxamide, Catocene
Curing agents	TDI(toluene diisocyanate), IPDI (isophorone diisocyanate), HMDI (hexamethylene diisocyanate)
Bonding agents	MAPO (tris-1-(2-methyl)aziridinyl phosphoric oxide, TEA (triethanol amino), MT-4 (adduct of 2 moles of MAPO, 0.7 mole adipic acid and 0.3 mole tartaric acid)
Combustion instability suppressants	Al, Zr, ZrC

Solid propellants can also be divided into three classes:

1.2.2.1.1. Double Base Propellants (DBPs) or Homogeneous Propellants

Double base propellants are conventionally known as homogeneous propellant. These propellants are based on cellulose nitrate (NC) and nitroglycerine (NG) mainly along with some additives like stabilizers, ballistic modifiers, plasticizers, cooling agents, lubricants, opacifiers etc.

1.2.2.1.2. Composite Solid Propellants (CSPs) or Heterogeneous Propellants

These propellants are heterogeneous and more energetic than DBPs. CSPs have high specific impulse, high flame temperature and high solid loading than DBPs. A comparison of DBPs and CPs are given in table 5.

Table 5. Comparison of Double Base and Composite Propellants

Properties	Double Base Propellants (DBPs)	Composite Solid Propellants (CSPs)
Type of propellants	Homogeneous	Heterogeneous
Specific impulse	Low (200-220s)	High (240-260 s)
Tensile strength	High (140 Kg/cm^2)	Low (15-25 Kg/cm^2)
Hazard classification	1.3	1.3
Density	Low (~1.6 g/cc)	High (~1.75 g/cc)
Shelf life	Longer (15-20 years)	Shorter (10-15 years)
Flame temperature	Low (2000-2500^{0}C)	High (2400-3200^{0}C)
Exhaust	Non-smoky	Smoky
Solid loading	Moderate	High
Degradation mechanism	NG exudes on ageing	Moisture sensitive

1.2.2.1.3. Composite Modified Double Base (CMDB) Propellants

In addition to conventional double base and composite propellants, composite modified double base propellants have been extensively used for military applications. Double base matrix of nitrocellulose (NC) and nitroglycerine (NG) is loaded with oxidizer like ammonium perchlorate (AP) or explosive ingredients like nitramines (RDX, HMX) to enhance the performance. Such loaded double base matrix constitutes CMBD class of propellants, where a gain of around 20-30 seconds in specific impulse is possible.

1.2.2.2. Liquid Propellants

The liquid propellants mainly consist of two components, oxidizer and fuel. Two pressurized tanks are used to hold oxidizer and fuel ingredients separately. Physical state of ingredients remains liquid and they are fed to combustion chamber through valves. A list of oxidizers and fuels along with their characteristics has been presented in table 6 and 7 respectively.

Table 6. Liquid Propellant Oxidisers

Oxidiser	Boiling point (^{0}C)	Melting point (^{0}C)	Density (g/cc)
Oxygen	-183.33	-217.78	1.14
Ozone	-111.11	-251.11	3.03
Hydrogen peroxide	152.22	-1.66	1.44
Nitrogen tetroxide	21.11	-9.44	1.45
Nitric oxide	-151.11	-161.11	1.27
Nitrous oxide	-89.44	-102.22	1.23
Nitric acid	86.11	-42.22	1.50
Tetranitromethane	125.55	12.78	1.65
Fluorine	-185.56	-223.33	1.11
Chlorine trifluoride	12.22	-83.33	1.17
Bromine pentafluoride	40.56	-61.67	2.46

Table 7. Liquid Propellant Fuels

Fuel*	Boiling point (^{0}C)	Density (g/cc)
Hydrogen (L)	113.51	0.07
Ammonia (L)	-252.75	0.82
Hydrazine	87.45	1.01
MMH	87.45	0.87
UDMH	62.85	0.791
TEA	89.45	0.723
Pentaborane	58.35	0.63
Aniline	184.35	1.03

* MMH = Monomethyl hydrazine
UDMH = Unsymmetrical monomethylhydrazine
TEA = Triethanol amine.

Different combinations of fuel and oxidizer are used to complete a liquid propellant. Some specific chemicals are also added to fuel and oxidizer to obtain the propellants of desired properties.

Liquid propellants can be classified in several ways. On the basis of boiling point liquid fuels are of two types:

i. *Earth Storable Fuel:* The liquid rocket fuel which can be stored in standard earth environment without phase change. Examples are hydrazine, monomethylhydrazine (MMH), aniline, unsymmetrical dimethylhydrazine (UDMH), pentaborane etc.
ii. *Cryogenic Fuel:* The liquid rocket fuel which need very low temperature working condition for their storage in liquid form. Their boiling points are less than standard atmospheric temperature on earth. They remain in gaseous form and are stored at high pressure. These fuels are highly corrosive and reactive in nature. Examples are hydrogen, ammonia etc.

1.2.2.3. Hybrid Propellants

Hybrid propellants are hybrid of solid and liquid propellants. They need solid fuel and liquid oxidizers. Ingredients of solid and liquid propellants are essentially adopted in hybrid propellants. For fuel rich propellants HTPB has been used as binder. Hybrid system can employ a variety of fuel-oxidiser combinations irrespective of their physical state and mutual compatibility. The solid propellant grains of hybrid propulsion have higher strength and resilience due to very low oxidiser content. This reduces creep and cracking of propellant grains. Hybrid charges are inert in nature and are safe to handle. Hybrid propulsion systems have low specific impulse and low fuel regression rates.

1.2.2.4. Gelled Propellants

Gelled propellant remains like thick paint in stationary condition but can flow through valves, pumps and other devices on application of adequate shear stresses. These propellants are basically a variant of liquid propellants which becomes thixotropic by some additives. In these gels, solid powdered ingredients can be added to enhance the density of the fuel and long term storage without setting or segregation. Gelled propellants are temperature sensitive which leads to change in viscosity, density and mixture ratio of propellant.

1.2.2.5. Fuel Rich Propellants (FRP)

Fuel rich propellants generally contain high percentage of fuel either metallic or polymeric. Mg/Al and their alloys, B, Zn, Ti, and Ni are promising metallic fuels. Among the polymeric materials HTPB and GAP can be used as fuel for FRPs. Recently; pyrolysable polymers like polycyclopentadienes are being attempted as fuel. Naphthalene / anthracene can also be used as a fuel for pressed fuel rich propellants.

1.2.3. Pyrotechnics

Pyrotechnics are composed of a mixture of a metal powder and an oxidizer. These are deflagrating low explosives, which find wide use in space applications for the manufacture of various devices such as squibs, chemical delays, catridges, igniter motors, etc. Boron, zirconium, titanium and aluminium are the metallic fuels and potassium nitrate, potassium perchlorate, ammonium perchlorate etc are the conventional oxidizers in these pyrotechnic compositions. However, ignition propagation characteristics under various conditions and output energy dictate the choice of pyrotechnic compositions.

1.2.3.1. Types of Pyrotechnics

Depending upon their functions, pyrotechnics are divided in five major categories-

1.2.3.1.1. Sound Producers

These are mainly used in the fireworks industry and in military simulators. These are further classified into two i.e. producer of a loud short duration sound, like bomb and grenade and a composition which produces shrill whistle of long duration. Typical sound producers are firecrackers, whistles, salutes and military grenade and ground – burst simulators.

1.2.3.1.2. Light Producers

These are basically of two types; flash powders and flares. The differences in the performance are in the bulk burning rates. Flash powders burn very rapidly and are used as time markers and lighting for night reconnaissance photography in military ammunitions.

Flares, on the other hand, burn very slowly and are employed to produce sustained lighting for a variety of purposes, including signaling and battlefield illumination. They are also used as warning flares that signals coming highway traffic during an accident or breakdown of a vehicle.

1.2.3.1.3. Heat Producers

Heat producing pyrotechnics are used for ignition mixes, thermites, sparks and incendiaries and heat pellets for thermally activated batteries. The ignition mixes are generally either metal / salt or metal / metal oxide mixes with no binder.

1.2.3.1.4. Smoke Producers

These are also subdivided into organic and inorganic types. The smoke produced by both types of pyrotechnics are used as markers, signals and tactically, as obscurants. They can be made in variety of colours and optical densities. The organic pyrotechnics are aerosol droplets of condensed organic dyes that are vaporized out of a relatively low temperature smoke mix.

1.2.3.1.5. Delay Producers

Delay producers are pyrotechnic devices consist of a mix that burn at a controlled rate to create a predetermine time delay between the ignition event and output. Pressed delay mixes utilize both metal/metal oxide and metal/salt formulations. Cord-type delays, such as safety fuse, quarry cord and igniter cord are composed of mixtures of black powder and diluents.

1.2.3.2. Factors Affecting the Reactions in Pyrotechnics

Pyrotechnics differ from explosives in that the reaction rates are very low. They differ from the propellants in that in most cases comparatively little gas is produced. In the pyrotechnic compositions generally no binders or plasticizers are used and often they are considered as porous materials. Thus, unlike explosives and propellants, which have burning rates (or detonation rates), that depends primarily on density, temperature and pressure, in pyrotechnics the convective heat transfer due to hot gas permeation into the reactant materials ahead of the reaction zone is important. The burning rate of pyrotechnics is therefore not only affected by pressure, density and temperature, but also by porosity, particle size, purity, homogeneity (degree of mixing) and stoichiometry (fuel/oxidizer ratio).

1.2.3.3. Applications of Metal complexes in Pyrotechnics

The substances used in the past in pyrotechnic composition such as heavy metal salts are now avoided since they lead to environmental and health problems. Some alternative metal complexes have been synthesized to replace the conventional oxidizer like lead azide, potassium nitrate etc in view of their inherent drawbacks such as friction sensitivity, lesser stability to moisture and incompatibility with other materials used in ammunition. Nickel hydrazine nitrate [5] has been prepared from available raw materials and found insensitive to impact, friction or electrostatic charge but more sensitive to flame [6]. It is found to be a suitable replacement for lead azide as an intermediate charge in commercial detonators. Also, the reaction liquor, unlike lead azide, does not contain any useless ions and thus makes possible the repeated use of the liquor in continuous preparation of this primary explosive. N, N-bis(2,4,6 – trinitrobenzoyl)lead hydrazine, synthesized from lead acetate and bis(2,4,6 – trinitrobenzoyl)hydrazine with better heat resistant properties is found to be a good substitute for lead styphnate [7]. Cobalt perchlorate complex with diaminourea was found to be good initiator for use as primary explosive [8].

1.3. THERMO-ANALYTICAL TECHNIQUES

Thermo-analytical techniques will be defined as those techniques in which some physico-chemical changes of the system is determined either as a dynamic function of temperature or measurements are made at a fixed or isothermal temperature. Using this definition the principal techniques of thermal analysis especially used for analysing HEMs are thermo-gravimetric analysis (TGA), differential thermal analysis (DTA) and differential scanning calorimetry (DSC) [9-13]. Other less widely employed but useful techniques include evolved gas detection or analysis (EGD or EGA) [14, 15], thermo-mechanical analysis (TMA) [16, 17], dynamic reflectance spectroscopy (DRS) [17, 18], electrical conductivity (EC) [19], photo-thermal analysis (PTA) [20, 21] and so on. Two other important thermal techniques that do not fall under the above classification are cryoscopic analysis [22, 23] and thermometric titrimetry [24, 25]. The various thermal analysis techniques are summarized in table 8.

Table 8. Various thermoanalytical techniques

S. No.	Technique	Abbreviation	Property	Uses
1.	Thermogravimetry (Thermogravimetric analysis)	TG TGA	Mass	Decompositions Dehydrations Oxidation
2.	Differential thermal analysis	DTA	Temperature difference	Phase changes reactions
3.	Differential scanning calorimetry	DSC	Power difference	Heat capacity Phase changes Reactions Calorimetry
4.	Thermomechanical analysis	TMA	Deformations	Meachanical changes Expansions
5.	Dynamic mechanical analysis	DMA	Moduli	Phase changes Polymer cure
6.	Dielectric thermal analysis	DETA	Permitivity	Phase changes Polymer changes
7.	Evolved gas analysis	EGA	Gases	Decompositions Catalyst and surface reactions
8.	Thermoptometry		Optical	Phase changes Surface reactions Colour changes
9.	Thermosonimetry	TS	Sound	Mechanical and chemical changes
10.	Thermomagnetometry	TM	Magnetic	Magnetic changes Curie points
11.	Thermoluminescence	TL	Light emitted	Trap depths
12.	Emanation thermal analysis	ETA	Gas released Stuctural changes Two or more techniques used on the same sample at the time	
13.	Simultaneous thermal analysis	STA		
14.	Controlled-rate thermal analysis	CRTA	The rate of change of the property is held constant	

Thermal techniques play a vital role in the research and development of various materials including HEMs. These techniques provide information about the thermal stability of parent compound, intermediate compound and final product. Most of these techniques can be employed effectively to follow the physical changes of state like polymorphic transformations, fusion, vaporization as well as chemical transformations involving thermal decomposition, oxidation and reduction reactions involving the mass changes. Some of the thermo-analytical techniques are employed to determine the storage life of various organic and inorganic compounds. Some basic aspects of the most common thermo-analytical techniques applied in enrgetic materials research are briefly described here.

1.3.1. Thermo-Gravimetric Analysis (TGA)

TGA is one of the most important thermal techniques and it measure mass change in a sample as a function of temperature, when it is subjected to a programmed temperature change [26, 27]. The plot of mass change versus temperature is termed as a thermo-gravimetric or a TG curve. The continuous recording of mass change offers special advantage; in that there is much less possibility of missing the step corresponding to the formation of the weakly stable intermediate in the multistep thermal decomposition reactions. Three modes of TG may be described:

a. Isothermal or static thermo-gravimetry, in which the sample mass is recorded as a function of time at constant temperature.
b. Quasistatic thermo-gravimetry, in which the sample is heated to a constant mass at each of a series of increasing temperatures: and
c. Dynamic thermo-gravimetry, in which the sample is heated in an environment whose temperature, is changing in a predetermined manner, preferably at a linear rate.

Resulting mass change versus temperature curve which has various synonyms such as thermolysis curve, pyrolysis curve, thermogram, thermogravimetric curve, thermogravigram, thermogravimetric analysis curve and so on, provides information concerning the thermal stability and composition of the initial sample, the thermal stability and compound formed of intermediate and the residue. Like other techniques, TGA has been refined with modern instrumentation. The part of the TG instrument used for monitering the mass change as a function of temperature is termed as thermobalance. Thermobalance consists of a balance and the furnace as main components. The balance moniters the mass change and the furnace heats the sample at the programmed rate or isothermally. The other important accessories of a thermobalance are the gas and vacuum manifolds, temperature programmer and sample carrier.

TGA is extensively used to determine changes in composition due to dehydration, decomposition and reaction with experimental atmosphere. It has also been employed for the determination of reaction kinetics. A small initial weight loss, $W\text{-}W_0$, generally results from desorption of a solvent or moisture. If this occurs within the vicinity of 100^0C, it is generally assumed to be associated with the evolution of water.

Subsequent calculations based on the thermal curve must be corrected for such a loss. The material being examined can be assumed to be stable upto temperature T_1 where it undergoes extensive decomposition with a total weight loss of $W_0\text{-}W_1$. The percent of weight lost, $(W_0\text{-}W_1/W_0)\times100$, can be used to determine the nature of decomposition (W_0 = initial weight, W_1 = weight at temperature T_1). Like any other technique, there are a number of operational parameters that affect the reliability and reproducibility of thermal curves. The major sources of error in TG appear to be due to temperature measurement, heating rate, aerodynamic forces, furnace atmosphere and heat of reaction studied.

Some characteristic temperatures called the inception, inflection and final temperature are generally assigned to the TG curve. Inception temperature indicates the onset of the mass change in the sample and generally corresponds to the temperature at which the cumulative mass change exceeds the sensitivity of the thermogravimetric system. The inflection temperature corresponds to the temperature at which the rate of mass change with respect to temperature is maximum. The final temperature represents the temperature at which the mass loss/gain of the given step is completed. These temperatures, for a given sample vary with several experimental parameters such as the sample size, heating rate, type of sample holder, ambient atmosphere etc. and sample characteristics such as particle size.

During TG measurement, two competing processes occur: those increasing the surface area, such as strains resulting from differences between the original material and its degradation products and the tendency for sintering to occur. At elevated temperature, the sintering process becomes more important. Synthetic air is usually employed in the thermobalance atmosphere because of the day-to-day variation of the humidity of the laboratory air. Controlled atmosphere of H_2O, CO_2, H_2 and N_2 are representative of many controlled atmosphere employed in TG.

1.3.1.1. Derivative Thermogravimetric Analysis (DTG)

The information that is available from a plot of mass change versus temperature or time is often not sufficient to solve a problem. Some additional information about the mass change can be obtained from a plot of differential mass change (dw/dt) versus time or temperature. The first derivative of the weight change is reffered as derivative thermogravimetry (DTG). The derivative curve may be obtained either from the TG curve by manual differentiation methods or by electronic differentiation of the TG signal. Accessory equipment is available for most thermo-balances so that the DTG curve can be easily recorded along-with the TG curve. It is useful for kinetic studies and for interpreting complex thermo-gravimetry curve. DTG yields a peak corresponding to each step involving a mass change. It is possible to obtain qualitative or quantitative information from DTG curve (The dynamic heating causes the mutual merging of two steps which are close to each other and make their detection impossible on the TG curve).

Advantages of DTG have been summarized by Erdey *et. al.* [28] as follow:

1. The curves may be obtained in conjunction with TG and DTA measurements.
2. The curves for DTA and DTG are comparable, but the results of the former method indicate even those changes of state that are not accompanied by loss in mass. The curve by the latter method is more reproducible.
3. While the curve for DTA extend over a wider temperature interval, due to subsequent warming of the material after reaction, the DTG measurements indicate exactly the temperatures of the beginning, the maximum rate and the end of the change.
4. On the TG curves, changes following each other very closely cannot be distinguished, as the corresponding stages coincide.
5. The DTG curves are exactly proportional to the derivatives of the TG curves; therefore the area under the curves gives the change in mass precisely. Accordingly DTG can give exact quantitative analysis.

6. The DTG method can be used for the investigation of materials that cannot be analysed by DTA.

The peak temperature of DTG represents the temperatures at which the rate of mass change is at a maximum.

1.3.2. Differential Thermal Analysis (DTA)

DTA is a technique by which the reaction of a material to temperature can be followed by observing the heat absorbed or liberated. DTA is especially suited for studies of structural changes within a solid at elevated temperatures. Heat effects associated with chemical or physical changes are measured as a function of temperature as the material is heated or cooled at a uniform rate. As the sample temperature is varied, it will undergo a variety of changes, each being accompanied by the release or absorption of energy. Melting, sublimation, phase changes, dehydration and boiling generally produce endothermic effects, while crystallization, oxidation and some decomposition produce exothermic effects.

In a typical apparatus, the temperature difference (ΔT) between the sample and the thermally inert reference is measured as a function of temperature when both are heated simultaneously at the programmed predetermined heating rate and a hypothetical DTA plot for the material undergoing a sequence of endothermic and exothermic transformation is shown in figure 1.

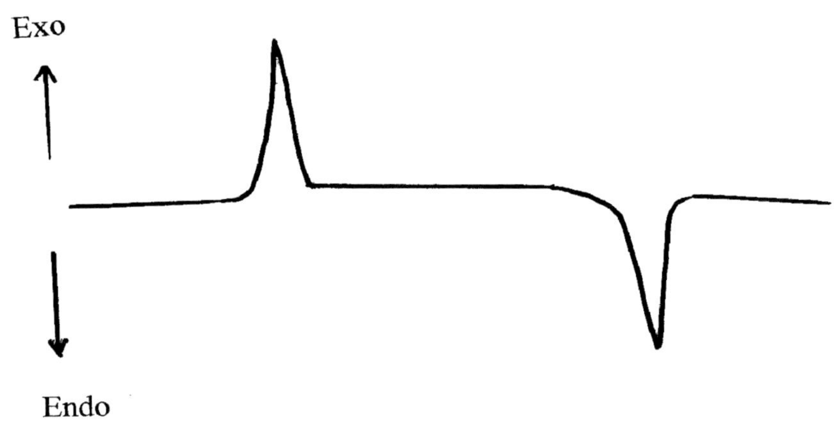

Figure 2. DTA plot for the material undergoing a sequence of endothermic and exothermic transformations.

The sample under investigation as well as the thermally inert reference is contained in identical sample holders made either of silica, alumina, platinum or platinum group metal alloys. The sample and thermally inert reference materials (like Alumina, MgO) contained in DTA sample holders are placed in the uniform temperature zone of the furnace, usually located in the center of the furnace and heated at the predetermined programmed rate.

Different types of thermocouples are employed in the fabrication of DTA instrument. The choice is dictated by the desired sensitively, the ambient atmosphere used in the experiment and the maximum temperature of operation. The thermocouples made of base metal alloys such as chromel- alumel are employed up to 1100^{0}C. For temperatures above 1100^{0}C the platinum-platinum 10 % rhodium or platinum- platinum 13 % rhodium, thermocouples are used.

Sensitivity of DTA is not limited by the type of thermocouple used but can be improved considerably by employing a suitable amplifier with sufficient gain. It is however not possible to amplify the ΔT signal indefinitely since along with signal, the noise is also amplified. In some of the latest designs, an alternate method is used in which the differential signal ΔT is amplified at source employing the thermopiles, which consist of several thermocouples connected in series.

1.3.2.1. Determination of Onset Temperature in DTA

The temperature of initiation of DTA peak as well as peak temperature is found to depend on the heating rate employed. In addition, other factors such as sample size also significantly influence these temperatures. The extrapolated onset temperature is found to be much less influenced by these factors. The 'extrapolated onset temperature' is defined as the point of intersection of the extrapolated base line of the DTA curve with the tangent drawn at the maximum slope on the rising part of the curve.

If any process takes place following the sequence of closely located steps than higher heating rate results in the overlap of the peaks. Slower heating rates facilitate resolution of peaks but results in the decrease in the peak size. Similar effect has been observed for sample size. Smaller samples yield better resolved peaks [29].

The two additional factors that need consideration in the interpretation of DTA peaks are 1) Ambient atmosphere and 2) Material of fabrication of the sample container. The former is particularly important when oxidizable gaseous species such as CO or NH_3 are evolved in the reaction which in presence of oxygen in air oxidize exothermically to CO_2 and (H_2O and NO_2) respectively. Crucibles made of platinum and alloys act as catalysts in the oxidation of CO to CO_2 and NH_3 to NO_2 in air. Silica and Alumina however does not possess this property. Simultaneous exothermic processes, taking place in the system, often dominate the endothermic decomposition process in such cases. Such type of oxidation reactions are dependent on the catalytic surface available during the process of evolution of gaseous products. Oxidation of ammonia released in the thermal decomposition of ammonium molybdate is facilitated by platinum crucible. If the reaction is carried out in Al_2O_3 crucible no exothermic effect is observed due to oxidation of evolved ammonia [30].

1.3.3. Differential Scanning Calorimetry (DSC)

Differential Scanning Calorimetry (DSC) invented by Watson and coworkers [31] provides all the information available from DTA but is much more sensitive and quantitative compared to DTA. The sensitivity of the DSC technique is two to three orders of magnitude higher compared to that of conventional DTA. The artifacts arising due to the use of large sample size in DTA are therefore, virtually eliminated in DSC and the data obtained is more quantitative.

The area under the DSC curve yields directly the enthalpy change involved in the physico-chemical changes. The applications may be divided roughly into two categories

a. Physical changes and measurements, such as melting, crystalline phase change, changes in liquid, liquid crystalline states and in polymers, phase diagrams, heat capacity and glass transitions, thermal conductivity and diffusivity and emissivity.
b. Chemical reactions such as dehydrations, decompositions, polymer curing, glass formation and oxidative attack.

The DSC curve consists of the plot of the difference in the rate of heat transport between the sample and the inert reference computed from the difference in temperature between them, plotted against time or temperature. Two types of DSC instruments are available commercially based on the way heat is measured. These are heat flux DSC and power compensation DSC.

1.3.3.1. Heat Flux DSC

The heat flux instrument works on the principle of DTA and is often looked upon as the quantitative DTA with improved sensitivity. The improved sensitivity is obtained employing the area type higher sensitivity thermocouples made from materials like chromel – constantan. During any phase transformation or chemical reaction the temperature of the sample differs from that of the reference and hence the rate of heat transport from the furnace to the sample will be different from that to the reference material and will depend on the temperatures of the sample and the reference. The conversion of temperature difference into heat flux, quantitative compensation for problem areas, such as temperature dependence of thermal transport and sensor sensitivity, are carefully built into associated software and hardware.

1.3.3.2. Power compensation DSC

The other type of DSC instrument is known as power compensation DSC. In this instrument the energy changes accompanying any physical or chemical transformation taking place in the sample are compensated in the reference so that the temperature difference between the sample and the reference is always maintained zero. The amount of energy needed to compensate the energy absorbed or released in the sample during the physical or chemical change gives directly the energy absorbed or evolved during the process. The area under the DSC curve gives directly the energy change involved in the process.

The DSC curve consists of the plot of the difference in the rate of heat transport between the sample and the inert reference compound, computed from the difference in temperature between them, plotted against time or temperature.

As in the case of DTA, the DSC curves are also influenced by a number of experimental variables and sample characteristics. It was assumed earlier that the thermal constant of the power compensated DSC was independent of temperature. This was however found to be far from being true. Thus, as in DTA the proportionality constant between the heat and area of the peak had to be determined at series of temperatures, if DSC has to be employed for the quantitative estimation of the energy changes accompanying the reaction.

In addition to the two DSC instruments described, there has been a recent innovation, which has resulted in the development of yet another more versatile DSC instrument [32]. This is termed as Modulated Differential Scanning Calorimetry (MDSC). This instrument is used to study the same material property changes, which are evaluated by conventional DSC including transition temperature, melting and crystallization and heat capacity. However,

MDSC provides unique capabilities, which increase the amount of information that can be obtained from the single experiment, thereby improving the quality of interpretation as well as productivity. These unique capabilities include:

1. Direct heat capacity measurement
2. Improved resolution of closely occurring and overlapping transitions
3. Increased sensitivity for suitable transitions
4. Separation of reversing and non-reversing phenomena

1.3.3.3. Factors Affecting the DSC Curve

The extensive use of DSC in the present book necessitates thorough knowledge about the various factors that may affect the accuracy and precision of the data obtained by DSC. Following are some of the factors, which can have considerable bearing on the data obtained.

a) Reference Materials:

In differential scanning calorimeter a suitable material should be used to balance the reference material and sample container with respect to weight or specific heat. The material should not undergo any interfering transition or adsorption of water.

b) Sample size
c) Atmosphere in the instrument (oxidizing or inert)
d) Heating rates and Calorimetric Sensitivity of the Instruments.
e) Temperature and Enthalpic calibration of DSC.
f) Impurities

The sample size, heating rate, impurities and type of sample container has a profound influence on the nature of the DSC curve. A typical example demonstrating the influence of heating rate on the peak resolution is provided in the study of phase transition in Temazepan [33]. The impurities in the sample influence the melting point of the pure compound. This fact can be useful for the separation of the overlapping transformations. The impurities need not necessarily be distinctly different chemical entity but could also be the polymorphic form of the parent compound. The presence of polymorphic form as an impurity can catalyze the transformation of the parent compound into the form of the polymorph existing as an impurity. Such influence of crystalline modifications has been studied for Gepirone hydrochloride [34].

The type of the sample pan used in DSC can have spectacular influence on the nature of the recorded curve. The extent of this influence is highlighted in a review by Giron [35]. It is interesting to note that for the given instrument the magnitude of the peak depends on the location of the sample in the sample pan, a fact which could otherwise be considered as trivial.

The DSC instrument has to be calibrated for temperature as well as enthalpy. This is done employing the standard materials recommended by International Confederation of Thermal Analysis and Calorimetry (ICTAC). An excellent monograph on DSC technique has been recently published by Hohne and co-workers[36]. Both qualitative and quantitative aspects of DSC including the instrument calibration have been covered exhaustively in this book.

DSC can be employed in all the applications where DTA can be used. The high sensitivity of DSC compared to DTA however, makes it more attractive for acquisition of the quantitative thermodynamic data such as heat capacity and enthalpy of formation.

1.3.5. Simultaneous or Hyphenated techniques

The coupling of TG and DTA was inevitable since the former shows the extent of loss in mass and the latter detects transformation involving energy changes. The coupling generally permits differentiation between physical and chemical changes. Many thermograms exhibit phenomenon occurring so closely in succession that assessment of significant temperature is difficult. This problem can be overcome by DTG (Differential thermogravimetric analysis), in which the rate of weight change is plotted against temperature.

1.4. Summary

Energetic materials are substances having high content of chemical energy. They contain two groups – explosophoric and auxoplosive groups. These can be used in several ways as explosives, PBXs, propellants and pyrotechnics etc. Thermal analysis can be effectively used to study the physico-chemical process that influence the safety and performance of energetic materials. The most commonly employed thermoanalytical techniques are TGA, DTA, DSC and their hyphenated techniques.

References

[1] M. A. Cook, The science of High Explosives, Reinhold Publishing Corporation, New York, 1963.
[2] J. A. Kent, Riegel's Hand Book of Industrial Chemistry, 7th Ed. Van Nostrand, Reinhold Co., New York, 1974.
[3] V. Pletis, *Zh. Obshch. Khim.*, 5 (1953) 173.
[4] C. B. Storm, J. R. Stine and J. F. Kramer, Chemistry and Physics of Energetic Materials, Kluwer Academic Publishers, Dordrecht, (1990) p. 605.
[5] Hariharanath, A.G. Rajendran, K.S. Chandrabhanu, M. Ravindran and C.B. Kartha, in: Proceedings of the 3rd International High Energy Materials Conference and Exhibit HEMCE – 2000, pp 469.
[6] Zhu Shun Guan, You Chen, Zhang Wen Yi and Mu Jing Yan, *Prop. Explos. Pyrot.*, 22 (1997) 36.
[7] D. Sheng and F. Ma, Huogonpin, 41 (1997) 36; CA 132 – 66235.
[8] S.A. Dehrle, *J. Energ. Mater.*, 15 (1997) 125.
[9] A. W. Coats and J. P. Redfern, *Analyst,* 88 (1963) 906.
[10] C. Duval, Inorganic Thermogravimetric Analysis, 2nd ed., Am. Elsevier, Newyork, 1963.
[11] P. D. Garn, *Anal. Chem,* 33 (1961) 1247.

[12] S. Gordon and C. Campbell, *Anal. Chem.,* 32, 271R (1960).

[13] Symposium on Thermal Methods of Analysis, *Anal. chem.,* 32 (1960) 1558; 35 (1963)184.

[14] W. Lodding, Gas Effluent Analysis, Marcel-Dekker, NewYork,1967.

[15] A. S. Kenyon, 'Techniques and methods of Polymer Evaluation, Marcel- Dekker, NewYork, 1966, chap. 5.

[16] W. W. Wendlandt, A*nal. Chem. Acta,* 33 (1965) 98.

[17] W. W. Wendlandt, *Science,* 140 (1963) 1085.

[18] Anon., *Chem. Eng. News,* (1963), 62.

[19] W. W. Wendlandt, *Thermochim. Acta,* 1 (1970) 11.

[20] D. J. David, *Thermochim. Acta,* 3 (1972) 277.

[21] G.N. Rupert, *Rev. Sci.Instr.,* 34 (1963) 1183.

[22] W. M. Smit, Z. *Elektrochem.,* 66 (1962) 779.

[23] E.M. Barrall and R.D. Diller, *Thermochim. Acta,* 1 (1970) 509.

[24] J. Jordan, Record of Chemical Progress, Hooker-Kresge Library, 19 (1958) 193.

[25] H. J. Keily and D.N. Hume, *Anal. chem.,* 28 (1953) 1294.

[26] W.W. Wendlant, "Thermal Methods of Analysis," Wiley- Interscience Publication, New York, Vol.19 (1974).

[27] J. Haines, Thermal methods, Applications and Problems, Publ. Blackie Academic and Proffessional, an Imprint of Chapman and Hall, Wester Cleddens Road, Bhishopbriggs, glasglow, G64 2NZ, (1995).

[28] L. Erdey, F. Paulik and J. Paulik, Nature, 174 (1954) 885.

[29] A. Blazek., "Thermal Analysis" Van Norstand, Reinhold Co., London, 5 (1973).

[30] G. Liptay, M. Berenyi, E. Sarkany, *Hung. Sci. Inatr.,* 15 (1969) 31.

[31] E.S. Watson, M.J.O.'Niel, J. Justin and K. Brenner, *Anal. Chem.,* 36 (1964)1233.

[32] D.R. Kamm, B.S. Crowe, M. Reading "Modulated Differential Scanning Calorimeter",Proc. Of 9th National Symposium on Thermal Analysis, Goa University, Eds. P.V. Ravinrdan, M. Sudersanan, S.R. Bhardawaj and S.R. Dharwadkar (1993) 19.

[33] C.W.H. Hohne, W. Hemminger and H. Hammershein, "Differential Scanning Calorimetry-An Introduction for Practitioners", Spinger Verlag, Berlin, 1966.

[34] R.J. Behrne, D. Brooks, R. F. Farney and T. T. Kensler, J. Pharm. Sci. 74 (1985) 1041.

[35] D. Giron, Thermochim. Acta. 248 (1995) 1.

Chapter 2

THERMOLYSIS OF AMMONIUM PERCHLORATE

ABSTRACT

Crystal structure of ammonium perchlorate (AP) has been discussed. The mechanism of thermolysis and conductivity of AP has been discussed thoroughly. Topography of the thermal decomposition has also been summarized. The radiolysis of AP has been described. Effects of various additives on thermolysis of AP are also discussed.

2.1. INTRODUCTION

Ammonium perchlorate (AP) is the 'work horse' oxidizer used in composite solid propellants for rocket and satellite launch vehicles. In the very beginning of study, the mechanism of thermal decomposition of AP was not clear. Now, there are at least three points of view from which the process of thermal decomposition starts. Most acceptable view is that the thermolysis of AP starts with proton transfer from cation to anion, similar to decomposition of all such ammonium salts. The nature of thermal decomposition of AP below phase transition at 240 ^{0}C and above this point is different. Low temperature decomposition (LTD) of AP is extremely sensitive to the preliminary mechanical treatment and irradiation, the presence of admixtures in it and the topography of a crystal. Addition of simple and complex metal oxides has a very strong effect on the rate of thermal decomposition. Below ~240 ^{0}C reaction arrest after the decomposition of almost one third of AP. High temperature decomposition (HTD) proceeds till completion; no dry residue remains.

2.2. CRYSTAL STRUCTURE

Ammonium perchlorate is a white crystalline substance. It exists as two important crystal modifications: one is orthorhombic which exist below 240 ^{0}C and other is cubic one which has existence above the temperature 240^{0}C. Solid state phase transition of orthorhombic to cubic phase takes place at a temperature of 240 ^{0}C. Heat of phase transition from orthorhombic to cubic modification is 11.3 kJ/mol [1]. The unit cell parameters are given in Table 1.

Table 1. Unit cell parameters of AP modifications

Modification	Unit cell parameter	Description
Orthorhombic (T<240 ^{0}C)	a = 0.9202 nm b = 0.5816 nm c = 0.7449 nm	Four oxygen atom form a tetrahedron around the chlorine atom. The distance between Cl and O is 0.143 nm. Each NH_4^+ ion is surrounded by 12 O-atoms. Eight of them have Cl-O distance 0.294 – 0.308 nm and four have 0.325 – 0.352 nm.
Cubic (T>240 ^{0}C)	A = b = c = 0.763 nm	Sodium chloride structure

The density of cubic modification (1.76 g/cm^3) is less than the orthorhombic structure (1.95 g/cm^3). NMR, neutron diffraction and ESR studies have shown that both the NH_4^+ and ClO_4^- ion can rotate freely [2,3]. In cubic structure both ions are free to rotate but in orthorhombic phase only ammonium ion rotates.

2.3. Dislocations in AP Crystal

The presence of dislocations in AP crystal was studied by means of etching method [4]. The etching agent may be ethanol or butanol or mixtures of these etchants. Etching carried out at rhombohedral and prismatic faces of AP crystal have indicated that the etch pits at the rhombohedral face corresponds to the screw dislocations emergent from surface and elongated along the longer axis of rhombohedron. In addition, it was found that dislocations are mainly grouped around the defects of the crystal growth. Etch pits are rectangular at the prismatic faces.

The initial dislocation structure of AP crystals depends on mechanical treatment and crystallization conditions. Maximal density of dislocations correspond to the sites which had been mechanically deformed preliminarily. Crystals of AP containing no dislocations can be grown from the solution of supersaturated conditions. The dislocation free crystals are called dislocation free whiskers.

2.4. Conductivity of AP

Conductivity of AP is due to the charge transfer through crystal lattice. There are three points of view on the mechanism of conductivity (charge transfer) in AP crystals:

2.4.1. Electron Transfer Mechanism

Assumption of electron conductivity in AP was made by Maycock and Verneker [5, 6]. The conductivity due to electron has been observed above 233 ^{0}C. However, no special confirmation of the existence of electron conductivity in AP crystal has been reported. The proof in the favour of electron transfer mechanism is the coincidence of the activation energy determined experimentally (5.6 eV) with the theoretical value (6.06 eV). The set of experimental data available at present make this assumption improbable.

2.4.2. Ion Transfer (by the Point Ion Defects) Mechanism

Zirkind and Freeman were the first to propose the ion mechanism of charge transfer in AP based on the work of Maycock and Verneker [7]. Ion transfer was assigned to low temperature region (90–150 ^{0}C). Above 150 ^{0}C charge carriers are vacancies which are Schottki ion pairs. There are two proofs in the favour of ion transfer mechanism of conductivity:

i. The activation energy of conductivity in AP and in Rubidium perchlorate is nearly same. Rubidium perchlorate undoubtedly exhibit ion transfer mechanism for conductivity.
ii. Orthorhombic modification of AP have similar crystal lattices of Rubidium perchlorate.

The above two data are insufficient to prove the mechanism of conductivity. Ion transfer mechanism can not explain the strong dependence of conductivity on the composition of gas phase atmosphere, especially NH_3 above AP during thermal decomposition etc. Ion transfer model of conductivity could not hold its position for a long time and gave place to proton transfer mechanism.

2.4.3. Proton Transfer Mechanism

Various experiments have proved that the mechanism of conductivity in AP is proton type [8, 9, and 10]. The mechanism of proton conductivity in AP proposed by Jacob can be summarized as-

i. Proton detachment from the ammonium ion.
ii. Transfer of proton to a trap and
iii. Transfer of formed proton vacancy along the cation lattice.

Another mechanism of proton conductivity was proposed by Khairetdinov [11]. It consists of three stages-

(i) Formation of current carriers by proton transfer from cation to anion.

$NH_4^+ + ClO_4 \rightarrow NH_3 + HClO_4$

(ii) Reorientation of protonated anion

$HClO_4 \rightarrow ClO_4H$

(iii) Proton jump from one anion to another

Thus, proton migration proceeds over the anion sub-lattice with reorientation due to rotation of ClO_4^- anion.

There are certain experiments summarized below that favours the proton conductivity in AP:

i. Doping of AP crystals with HSO_4^-, which is a proton donor, causes an increase in conductivity and decrease in activation energy.
ii. Jacob showed that the conductivity of AP changes by doping with Ba^{++} and SO_4^{--}, the effect of water vapour, ammonia and cycling for conductivity. The formation of ammoniate in presence of gaseous ammonia can be the reason of large increase in conductivity of solid AP.
iii. G. Singh *et al* have also considered importance of proton transfer for thermolysis of alkyl and aryl substituted ammonium salts including ammonium perchlorate [12].

2.5. Thermal Properties of AP

Ammonium perchlorate is considered to be a rather stable chemical compound. It starts to decompose at 130 ^{0}C. Decomposition of AP is about 10 times slower than ammonium nitrate at 130 ^{0}C. However, the thermal stability of AP depends on the method of crystallization, preliminary treatment and storage time after obtaining the crystals. Crystalline AP exists in the orthorhombic form below 240 ^{0}C. At 240 ^{0}C it changes to cubic form. Some data on various thermal properties of AP is given in table .

Table 2. Data on thermal properties of AP

Property	Value
Phase transition temperature	240 ^{0}C
Melting temperature	560 ^{0}C
Surface temperature	$T_s = f(P)$
Flame temperature	1127 ^{0}C
Heat of phase transition	20 cal g^{-1}
Heat of sublimation	496 cal g^{-1}
Heat of decomposition	-265 cal g^{-1}
Heat of melting	60 cal g^{-1}
Specific heats of	
a) Orthorhombic solid	0.34 cal g^{-1} C^{-1}
b) Cubic solid	0.34 cal g^{-1} C^{-1}
c) Melt	0.33 cal g^{-1} C^{-1}
d) Gaseous products	0.30 cal g^{-1} C^{-1}

2.6. THERMAL DECOMPOSITION OF AP

A typical TG-DSC thermal curve of AP under inert atmosphere is shown in figure 1.

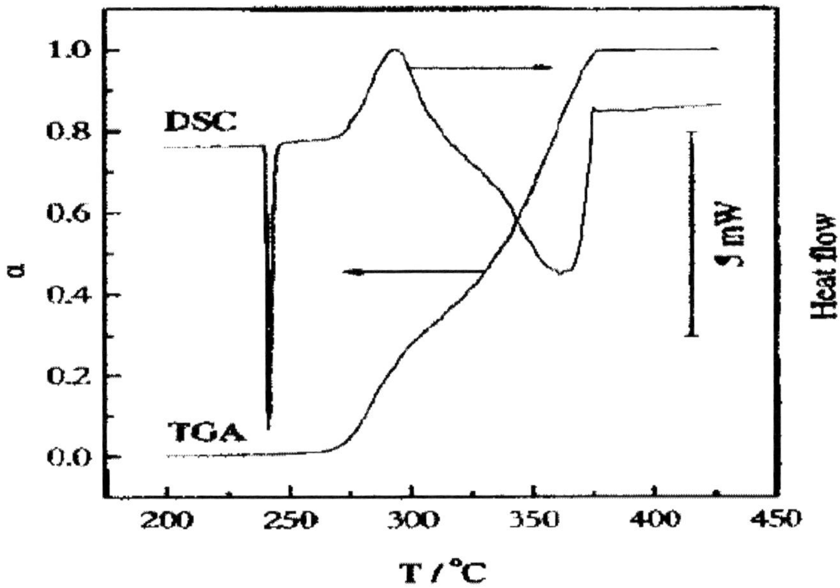

Figure 1. TG-DSC thermal curve of ammonium perchlorate under inert atmosphere at a heating rate of 5 ^{0}C min^{-1}.

A sharp endotherm is obtained for the solid state polymorphic transition from orthorhombic to cubic. Among many peculiar features of AP thermolysis, the most striking one is that there is a low temperature decomposition (LTD) below ~300 ^{0}C and resulting only in ~30% mass loss. The high temperature decomposition (HTD) is observed above 300 ^{0}C and leads to complete gasification of AP.

2.7. THERMAL DECOMPOSITION PRODUCTS OF AP

Dode was the first to report the decomposition products of AP [13, 14]. Products of thermal decomposition of AP depends on temperature. Low temperature decomposition (below 300^{0}C) proceeds according to the scheme

$4NH_4ClO_4 \rightarrow 2Cl_2 + 2N_2O + 3O_2 + 8H_2O$

In addition to the above products ClO_2, HCl and N_2 are formed.

High temperature decomposition proceeds as under

$2\ NH_4ClO_4 \rightarrow Cl_2 + 2NO + O_2 + 4H_2O$

In addition, traces of HCl, NOCl and NO_2 are also formed.

The above products were also confirmed by gas chromatography and mass spectrometric analysis. ClO_2 which is formed among the products of low temperature thermal decomposition was thought to initiate the combustion process. ClO_2 is a stable paramagnetic molecule and can be easily detected with the help of electron paramagnetic resonance.

In the above investigation the possibility of the effect of secondary processes occurring in the gas phase between the products formed during thermal decomposition has not been omitted by carefullly chosing the experimental conditions. Possibility of secondary processes can be excluded by keeping the sample near the ion source of high resolution mass spectrometer or by decomposing AP under high vacuum. Under these conditions, the products of decomposition were detected to be NH_3, HCl, Cl_2, O_2 and small amounts of nitrogen oxides. Analysis of the results obtained indicated that the primary products of thermal decomposition of AP are ammonia and perchloric acid. However, although gas phase secondary processes are excluded, the presence of $HClO_4$, NO, NO_2, O_2 and Cl_2 among the products under high vacuum indicated the occurrence of a parallel reaction proceeding in the condensed phase.

In a way, in the decomposition products, the product from two processes may also be included:

a. The possibility of superposition of sublimation processes on thermal decomposition processes and
b. The possibility of secondary reactions in pores between the crystals of AP.

Methods have been developed which avoid the loss of primary products (main products) in secondary reactions between them. With the help of this procedure, mass spectrometric data on thermal decomposition of AP at 400 ^{0}C, are shown in table 3.

Table 3. Mass spectrometric data on thermal decomposition of AP at 400 ^{0}C

m/e	Intensity
16 NH_2^+	53
17 NH_3^+, OH^+	93
35 Cl^+	13
51 ClO^+	13
67 ClO_2^+	73
83 ClO_3^+	100
100 $HClO_4^+$	53

This reported mass intensity ratio 35:51:67:83:100 is close to the ratio of the masses in perchloric acid decomposition. Thus, we may say that Cl^+, ClO^+, ClO_2^+ and ClO_3^+ are the fragmentation products of $HClO_4^+$.

AP dissociates into $HClO_4$ and NH_3 according to proton transfer mechanism *i.e.* $HClO_4$ and NH_3 are the primary products of thermal decomposition. Since, there is no time for the secondary reactions between main products, $HClO_4$ and NH_3 enters into mass spectrometer and give signals of NH_2^+, NH_3^+, OH^+, ClO^+, ClO_2^+, ClO_3^+ and $HClO_4^+$ by the fragmentation.

G. L. Pellet studied the thermal decomposition of AP under rapid heating with CO_2 laser [15]. The data shows that the primary products of pyrolysis are NH_3 and $HClO_4$. Moreover, the ratio of NH_3 and $HClO_4$ decreases permanently which is an evidence that NH_3 is removed more easily and $HClO_4$ is accumulated in the crystal. After 100 – 200 ms, H_2O, O_2, Cl_2, HCl,

ClO_2, N_2O and N_2 appears as the products of secondary reactions which take place on crystal surface. After 200 ms NH_3 and $HClO_4$ disappears gradually. The formation of NH_3 and $HClO_4$ as the primary product of thermolysis of AP can be considered as the confirmation of the idea of proton transfer mechanism of decomposition of AP.

It has already been discussed that low temperature decomposition (below 240 ^{0}C) of AP proceeds only upto 30% by mass. Three reasons have been proposed for the decomposition arrest:

i. Strong disordering of the lattice occurring during the reaction that this becomes an essential obstacle for electron transfer from anion to cation.
ii. At low temperatures, the most active mozaic structure is burnt away.
iii. Studies with optical and electron microscopy have shown that unreacted AP contains micro-sized pores. $HClO_4$ and NH_3 accumulate in these pores. Due to accumulation of $HClO_4$ proton equilibrium shift to left hand side. Also, selectivity of the pores for the removal of gaseous products and accumulation of water stabilizes the thermal decomposition of AP.

2.8. Topography of the Thermal Decomposition Process of AP

Thermal decomposition of AP starts not over the whole crystal volume or on whole surface. It starts in some sites near the outer and inner surfaces of a crystal. The starting points are called as active centers. These active centers are formed at dislocations. In other words, thermal decomposition of AP starts at dislocations being called as active centers. The reaction propagates near active centers with the formation and growth of reaction nuclei. Reaction nuclei grow and get merged to each other forming the reaction region.

Reaction nuclei are usually formed not at the very surface but at some distance form it, inside the crystal at a depth of 20-30μm. The nuclei are not continuous; they are separate seeds of about one micron in size. The shapes of nuclei are different at different faces of the crystal and depend on the temperature of decomposition. Below 240 ^{0}C (phase transition point), the nuclei at rhombic face of the crystal are cigar shaped and at prismatic face are rounded. Above 240 ^{0}C, the nuclei are spherical. Shape of nuclei also depend upon the crystal chemical features of a given face. For example; on rhombic face, the nuclei have a shape of rhomb, while on prismatic face they are rectangular. During decomposition, nuclei increases in size upto several microns and stop growing, their shape changes and gradually become round.

The positions of nuclei with respect to each other are different for different faces. The nuclei are in the movement condition along the diagonal of the rhomb at a velocity of 7-10 μm per minute (at 230 ^{0}C).

According to Raevsky each nucleus formed during thermal decomposition of AP is composed of three zones [16]:

i. the region of a porous core formed by junction of seeds
ii. the region of chemical reaction and

iii. the periphery of the growing nucleus

Thus, chemical reactions of thermal decomposition proceed in the region (ii) which starts mainly on dislocations.

The gaseous products evolved during the thermolysis into the sub-surface pores can raise pressure upto 20 atm, the crystal of AP acquires mechanical strain and gets deformed in this zone. Due to this, existing dislocation multiplicates and along the periphery generates new nucleus.

Reaction centers formed during low temperature decomposition of AP are partial decomposition centers. Inspite of these centers, Raevsky discovered also other centers and call them centers of complete decomposition. These centers are filled with liquid and do not stop growing once the process runs. According to Raevsky the liquid phase is an eutectic mixture of ammonium perchlorate with the products of its decomposition. But Galway thinks that the liquid phase is nitrile perchlorate, NO_2ClO_4, which is formed as an intermediate product resulting from oxidation of ammonium ion with the oxygen containing compounds of chlorine [17].

Completeness of overall thermal decomposition depends on the kinds of centers on which it proceeds. If the centers are of first kind, decomposition of AP proceeds only by 30% and the product of decomposition is porous perchlorate identical in its chemical composition to the initial compound. If decomposition occurs at centers of complete decomposition, thermolysis proceeds till completeness.

2.9. Growth and Rate of Nuclei Formation During LTD (upto 240^0C) of AP

Thermal decomposition of AP has been described by the use of different equations based on different models. Use of these models yields a single value of activation energy (E) for thermal decomposition. For such a solid state multistage thermolysis single value of activation energy has no sense. This may be illustrated by the dependence of E versus transformation degree plotted according to the data of different authors. One can see that E values can be different at different stages of the process. This method allows determining not the overall reaction rate but specific rate of the process. If this is unknown, it is very difficult to carry out a correct kinetic analysis of the heterogeneous processes. Kinetic analysis using this method have given an indirect indication that the process of thermal decomposition is not localized on the flat boundary between non-reacted part of a crystal and a nucleus but proceeds within some diffuse layer. The addition of homophase heterovalent additives (divalent cation) in the lattice of AP, cause a decrease in the induction period, an increase in the rate of nuclei formation and maximal number of nuclei per unit surface. However, the additives do not affect the rate of nucleus growth in lengthwise and crosswise directions.

The rate of nuclei formation is not affected by water vapour concentration but the rate of growth of nucleus is affected. However, the character of anisotropy of the nucleus development is unaffected. Synchronous changes occur in the growth rate of nuclei both in the direction parallel to the principal axis of symmetry and perpendicular to it.

Figure 2 shows the dependencies of the rate of nucleus formation, maximal number of nuclei (n_{max}) and the induction period on partial pressure of ammonia and perchloric acid. It is clear that an increase in the partial pressure of ammonia causes an increase in the induction period, a decrease in the rate of nuclei formation and in the number of nuclii. An increase in perchloric acid pressure leads to an increase in the nuclei formation rate. Partial pressure of ammonia also affects the nuclei growth rate, which decrease as partial pressure of ammonia increases.

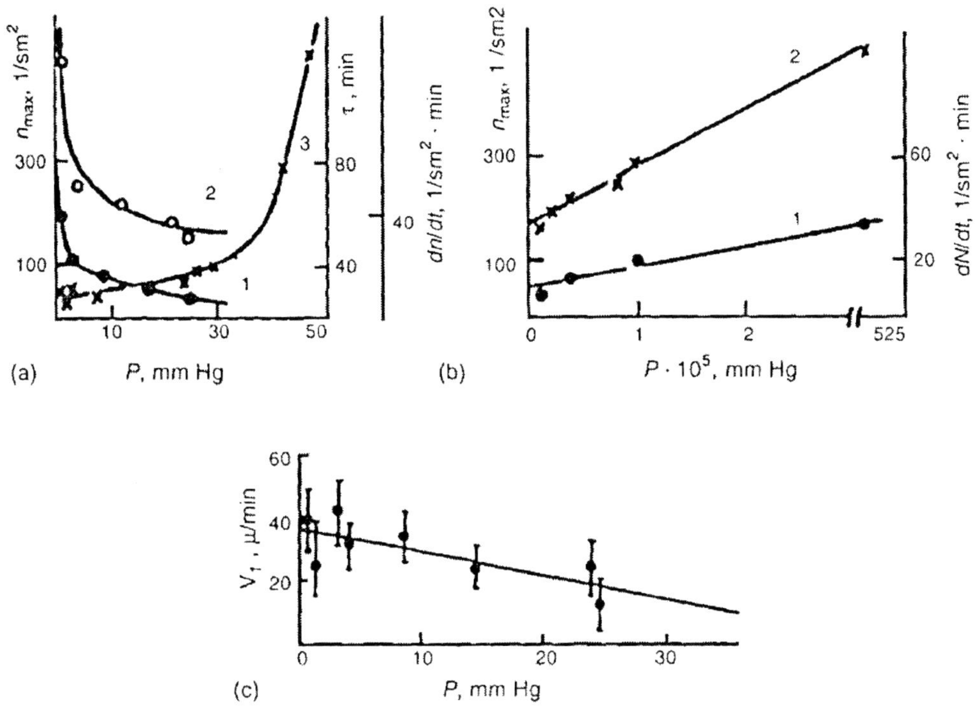

Figure 2. (a) Effect of partial pressure of ammonia for thermolysis of AP on 1 - rate of nuclei formation dN/dτ, 2 - n max and 3 - induction period τ; (b) Effect of equilibrium pressure of perchloric acid for thermolysis of AP on; - rate of nuclei formation dN/dτ and 2 - n max; (c) Effect of partial pressure of ammonia on the rate of reaction nuclei growth for thermolysis of AP.

2.10. Mechanism of AP Decomposition

Of the many mechanistic pathways for thermal decomposition of AP, the most important are described here in brief:

2.10.1. Electron Transfer Mechanism

One of the first versions of such a mechanism was proposed by Galway and Jacob [18]. According to them:

(i) at first electron transfer from perchlorate ion to ammonium ion occurs on the surface not in volume. Due to distance factor, an efficient acceptor of electrons will not be any ammonium ion but only those located in interstices.

$$ClO_4^- + NH_4^+ \rightarrow ClO_4^\circ + NH_4^\circ$$

(ii) After capturing an electron, the ammonium radical decomposes into ammonia and hydrogen atom.

$$NH_4^\circ \rightarrow NH_3 + H^\circ$$

(iii) Hydrogen migrates over the lattice. Electron migrates exactly in the same manner over the anion sub lattice. As a result of the interaction between ClO_4° radical and H, $HClO_4$ is formed. $HClO_4$ may continue interacting with H.

$$ClO_4^\circ + ClO_4^- \rightarrow ClO_4^- + ClO_4^\circ$$
$$H^\circ + ClO_4^\circ \rightarrow HClO_4$$
$$HClO_4 + H^\circ \rightarrow H_2O + ClO_3^\circ$$

(iv) ClO_3° radical is trapped by electron and transferred into ClO_3^- ion.

$$ClO_3^\circ + e \rightarrow ClO_3^-$$

(v) After that, ClO_3^- and ClO_4^- radicals can decompose, interact with NH_4^+ ions etc. As a result of interaction secondary products are formed; the major ones are Cl_2, N_2O and H_2O.

According to Raevsky and Manelis electron transition occur from valence band into conduction band. But ammonium perchlorate is a typical dielectric with the forbidden gap width about 5.6 eV, while kT value for decomposition temperature (450 ^{0}C) is about 0.2, it is evident that probability of such a direct electron transfer is low.

Galway and Jacob postulated the formation of molecular complex [NH_4ClO_4] due to local electron transfer as proposed by Bircomshaw [19]. The complex is transferred into final products. Formation of ClO_4 radical has been considered to be the limiting stage of nucleus formation process whereas the decomposition of molecular complex as limiting stage of nucleus growth. At first, water is formed during decomposition and then ClO_2 and N_2.

A very important feature of electron transfer mechanism is the requirement of a fresh surface for the decomposition to proceed. The Frenkel defects do not lead to the production of a fresh surface and hence the decomposition terminates at 30%. The defect structure in the temperature range 250-350 ^{0}C is of Schottky form. The creation of new Schottky pair produces a new surface and hence decomposition proceeds to completion.

2.10.2. Proton Transfer Mechanism

Bircomshaw and Newman [19], Jacobs and Whitehead [20] proposed proton transfer from cation to anion, resulting in adsorbed ammonia and perchloric acid on the surface of AP, followed by evaporation of ammonia and perchloric acid into gas phase, depending upon the prevailing temperature and pressure conditions, as shown in scheme 1. The decomposition according to proton transfer mechanism could be a direct bimolecular reaction between NH_3 and $HClO_4$. This notion is supported by the fact that decomposition of AP is accelerated by $HClO_4$ and retarded by NH_3.

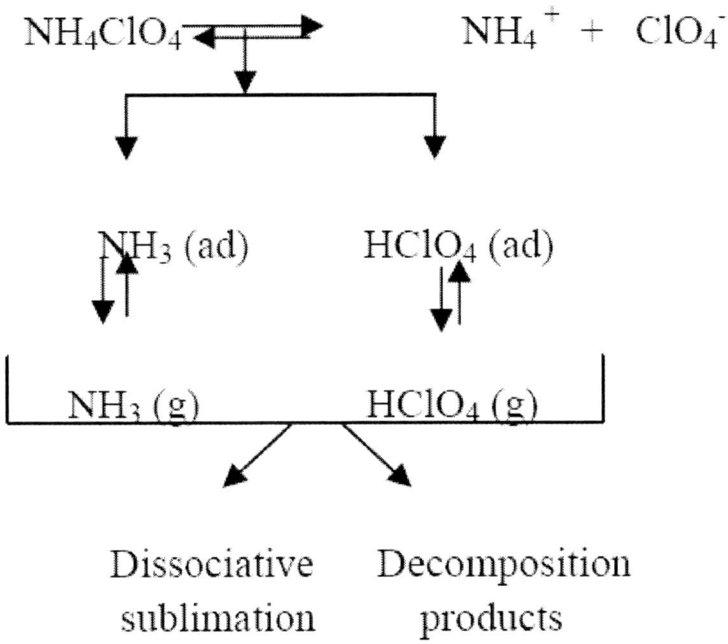

Scheme 1. Proton transfer mechanism of AP decomposition.

According to Jacob, at first molecular complex $[NH_3.HClO_4]$ is formed; then it decomposes into ammonia and perchloric acid. The molecules of ammonia and perchloric acid either react in the adsorbed layer on the surface of perchlorate or desorbs and sublime interacting in the gas phase.

At low temperature (<350 ^{0}C), the reaction of the surface proceeds more rapidly than sublimation. Above 350 ^{0}C, inverse relation of these processes is observed.

Thermal decomposition of AP proceeds through the formation and growth of reaction nuclei. The nuclei which are cavities filled with the gaseous products of decomposition of AP (centres of partial decomposition according to Raevsky) or with liquid (centres of complete decomposition) are formed as a result of proton transfer and capture by a trap. The sites for the occurrence of these processes can be linear defects of the AP lattice, e.g. dislocations. The efficient proton trap is ClO_3^- ion.

$$NH_4^+ + ClO_4^- \rightarrow NH_3 + HClO_4$$

$$HClO_4 + ClO_3^- \rightarrow HClO_3 + ClO_4^-$$

The formed $HClO_3$ is unstable and decomposes. Its decomposition products include ClO_2 which can interact with AP. It was demonstrsted that the interaction proceeds mainly as a result of the reaction with NH_4^+ ion. In addition, ClO_2 can enter chemical interaction with ammonia formed during the dissociation of AP.

$$ClO_2 + NH_4^+ \rightarrow H_2O + N_2O + HCl + R_1$$

$$ClO_2 + NH_3 \rightarrow N_2 + H_2O + Cl_2 + R_2$$

where R_1 and R_2 are the products of ammonia oxidation, which can be hydrazonium or hydroxyl ammonium ions, NO_2ClO_2 and so on.

As a result of these processes, the nucleus grows upto the size of 2-3μm. Then it stops growing. The gaseous products accumulate in pores which lead to an increase in pressure. The crystal around the nucleus gets deformed because of the pressure of the order of 20 atm inside the cavity. A consequence of deformation can be the formation of new dislocations.

It was shown that the high temperature thermal decomposition (>350 ^{0}C) is dissociation of AP into NH_3 and $HClO_4$, *i.e.* starts from the same elementary stage of proton transfer from NH_4^+ to ClO_4^- ion. The difference is that the process starts not in a sub-surface site but on the surface. Secondary chemical reactions between the products of thermal decomposition of NH_3 and $HClO_4$ occur either on the surface of AP or in the gas phase above the surface. If the conditions allow avoiding secondary reactions, for example when the process is carried out in vacuum, sublimation of AP occurs.

The mechanism of secondary reactions above the surface of AP changes with temperature and pressure. For example, when temperature rise, the NO to NO_2 ratio changes due to the decomposition of NO_2 at high temperature and due to possibility of direct interaction between NH_3 and $HClO_4$

2.11. Radiolysis of AP

According to the results of various investigations of the mechanism of the process occurring during the radiolysis of AP, excitation of ions occur under the action of irradiation both in the cation and anion sub lattices. The main effects are:

(i) Following are the possible processes in ClO_4^- ion under excitation [21- 23]:

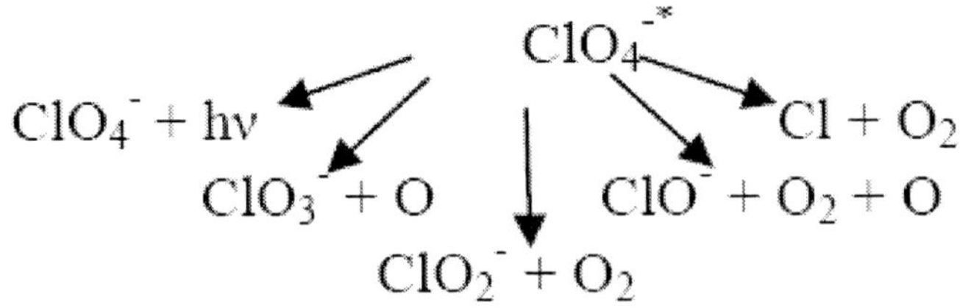

Ionizing radiation can cause ionization of ClO_4^- ion in addition to excitation:

$$ClO_4^- \rightarrow ClO_4^{\circ} + e \rightarrow$$

ClO_4° can either recombine with electron or further decompose.

$ClO_4^{\circ} \rightarrow ClO_3^{\circ} + O \rightarrow$

(ii) Ionization and excitation of the cation leads to the formation of NH_3^+ radicals.

$(NH_4^+)^{\bullet} \rightarrow NH_3^+ + H^{\circ} \rightarrow$

NH_3^+ radical is rather stable and starts perishing at a temperature of 300-370 ^{0}C. The concentration of ClO_3 radical is about two orders of magnitude lower than NH_3^+, because former recombine more rapidly.

Some of the products of radiolysis, e.g. ClO_3^-, ClO^-, ClO_2 and NH_3^+ were detected in the irradiated AP by means of chemical analysis, electron paramagnetic resonance and thermo-luminescence.

One of the most important results of the radiolysis is the accumulation of ClO_3^- ions in the crystal of AP. ClO_3^- ions are one of the most efficient traps for proton regenerating during the process. Another result of radiolysis which is important for subsequent thermolysis was considered to be the formation of additional number of dislocations and/or displacement of dislocations in crystal, starting from the adsorbed dose of the order to 10^5 radiation and higher. Some of the chemical and physical results of the radiolysis can lead and do lead to the acceleration of thermal decomposition process. Photolysis of AP corresponds in general to the scheme of radiolysis.

2.12. The Effect of Irradiation on Thermal Decomposition of AP

Some important effects of radiolysis on thermal decomposition of AP are discussed below:

(i) Under isothermal condition the induction period decreases after irradiation. When induction period of thermal decomposition of AP can reach several hundred or thousand hours, irradiation may effect even thermal decomposition at very low temperature (145-140 ^{0}C). Under these conditions samples composed of coarse crystals turn out to be most sensitive than grinded sample (figure 3).

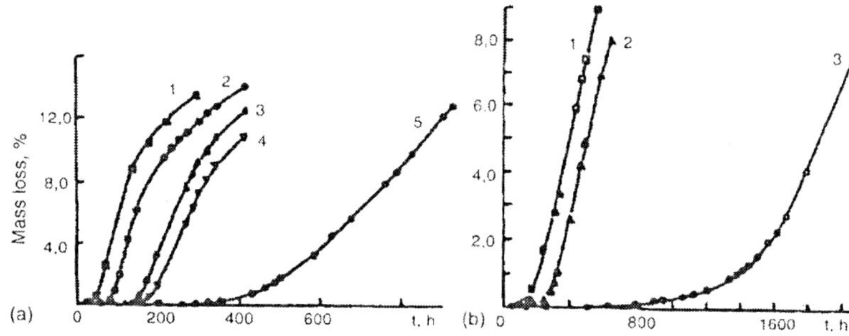

Figure 3. Kinetic curves of decomposition of AP at 135^{0}C.

(a) The sample not treated mechanically: (1) 300s, (2) 100s, (3) 60s, (4) 30s irradiated with

γ- rays and (5) non-irradiated sample

(b) The sample grinded in a mortar: (1) 300s, (2) 60s irradiated with γ- rays and

(3) non-irradiated sample

(ii) Freeman and Anderson [21, 22] showed that thermogravimetric and DTA curves of AP irradiated with X-rays and gamma radiation substantially differ from the corresponding curves of non-irradiated AP. An increase in the dose of X-ray radiation absorbed by AP inspires to proceed the thermal decomposition with a noticeable rate even below phase transition point (figure 4). in addition irradiation causes the appearance of a new peak at 360 ^{0}C and a shift of the peak at 450 ^{0}C, corresponding to high temperature decomposition.

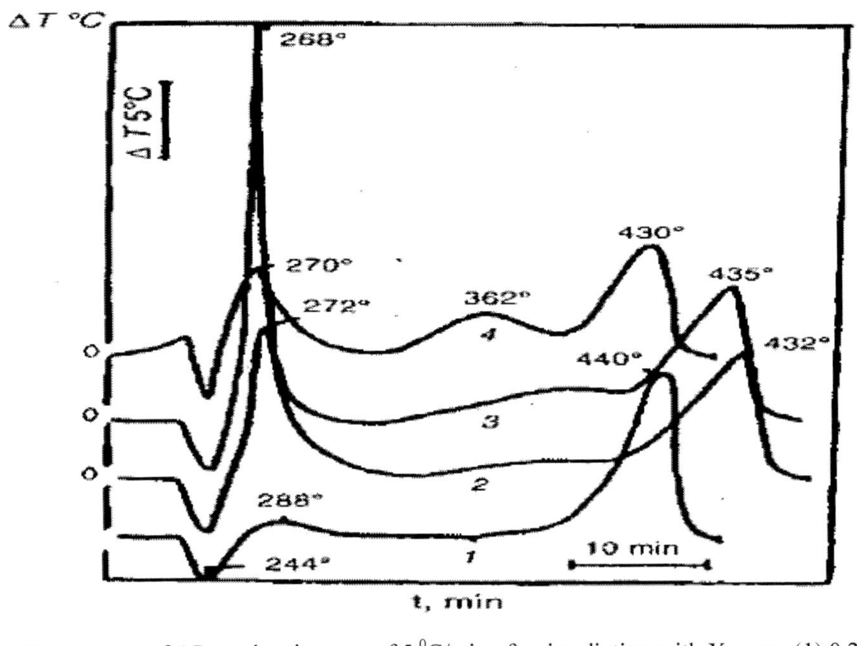

Figure 4. DTA curves of AP at a heating rate of 5 $^{0}C/min$ after irradiation with X-rays; (1) 0.27 eV/molec. dose, (2) 3.5×10^{-4} eV/molec. Dose; (3) 5.4×10^{-5} eV/molec. dose (4) non-irradiated sample.

(iii) Irradiation of AP with protons, fast neutrons or gamma radiation showed that most effective radiation is fast neutrons and protons and they accompanied by the formation of new peaks. Irradiation with gamma rays is less effective.

(iv) It was shown that irradiation with a dose upto 0.6 M rad causes more than a two fold decrease in activation energy. Irradiation by X-ray (dose rate 300 rad/min) during thermal decomposition of AP has no effect on the rate of thermal decomposition [23- 27].

(v) Irradiation of AP causes chemical and physical changes which lead to change in thermal decomposition pattern. There are two main reasons which affect the thermolysis of AP after irradiation:

(a) Chemical reason

Chemical analysis indicates that the chlorate ions accumulate in the crystals during irradiation. Chlorate ion is an efficient proton trap and it accelerate the thermal decomposition of AP. Argument in favour of this reason has been given by Freeman [21, 22]. He obtained same derivatographic curves to that of irradiated AP and AP containing chlorate ion introduced into crystal.

However, it has been showed that the effect of radiation treatment is higher than the doping of AP crystal with chlorate ions in the amount equal to that formed during radiolysis. So, it was assumed that, in addition to the formation of chlorate ion, the rate of thermal decomposition is also affected by another factor causing an increase in the reactivity of AP.

(b) Physical reason

Herley and Levy discovered that irradiation of AP causes dislocation displacement (figure 5a) [28]. On the other hand, if the dose absorbed by the crystal exceeds 105 rad, an increase in the number of dislocations occur (figure 5b) due to deformation of the walls of a pore in which radiolysis occurs. Since, dislocations are known to be the sites where the reaction nuclei are formed, an increase in the number of dislocations became the second reason of the influence of preliminary irradiation on the rate of thermal decomposition of AP

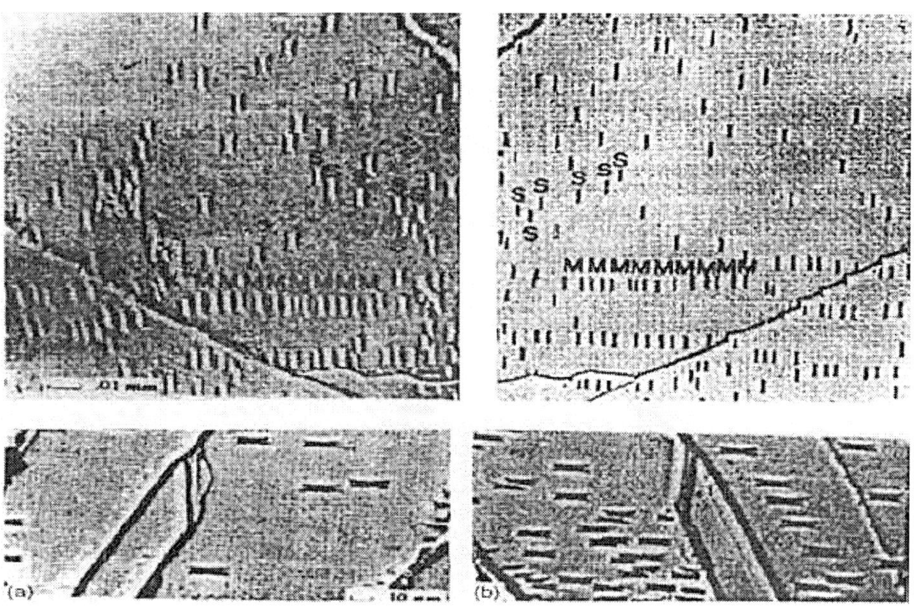

Figure 5. Displacement and generation of new dislocations due to irradiation with γ-rays; ('M' are marked for shifted dislocations, 'S' are marked for not shifted dislocations).

Experiments [28] have indicated that the main factor responsible for the influence of preliminary irradiation on the rate of thermal decomposition of AP is the formation of chlorate ions (proton traps) in the lattice under the action of radiation. Increase in dislocations in the crystal affect the thermal decomposition process, but this seems to be of secondary character.

(vi) Photolysis of AP corresponds in general to the scheme of radiolysis.

(vii) Under irradiation, potential centres arise over whole crystal volume. However, for these potential centres to transform into active centres, the conditions are likely to be realized in the sub surface layer of AP crystal and can not be realized in its volume due to cage effect.

(vii) Sensitivity of AP to irradiation can be changed by additives, which are the donor (sensitizers) and proton acceptors (inhibitors).

2.13. The Effect of Additives on Thermal Decomposition of AP

Thermal decomposition of AP is highly sensitive to the action of various additives. Depending upon the mixing of additives with AP, additives are classified into two clases:

(a) Homophase additives

These are the additives which enters into the lattice of AP. This is done by means of co-crystallization.

(b) Heterophase additives

These are the additives which form their own phase. This is done by mechanical mixing of AP and additives.

Additives can accelerate or decelerate the process of thermal decomposition (mainly LTD), which is important for the storage of AP and mix compositions of AP, such as propellants etc. The additives may also affect the deflagration delay time. For simplification, the effects of additives on thermal decomposition of AP would be discussed in four classes:

2.13.1 Effect of heterophase additives
2.13.2 Effect of homophase additives
2.13.3 Effect of heterophase precursor of the catalyst
2.13.4 Effect of homophase precursor of the catalyst

2.13.1 Effect of heterophase additives

Several transition metal oxides are known to accelerate the thermal decomposition of AP. Some of these oxides are Cr_2O_3, MnO, MnO_2, Fe_2O_3, CoO, Co_2O_3, NiO, Cu_2O, CuO, ZnO and CdO etc. Some non-transition metal oxides such as MgO, SiO_2 and oxides of rare earths also accelerate thermal decomposition of AP. Some transition metal fluorides are also reported to act as catalyst for AP thermolysis at 230^0C [29]. Other conditions being kept constant, the catalytic activity of fluorides are always lower than the corresponding oxides. The catalytic activity of fluorides changes in the sequence $Co^{++} > Mn^{++} > Zn^{++} > Cu^{++} > Ni^{++}$.

Spinels such as $NiCr_2O_4$, $NiFeCrO_4$, $NiFe_2O_4$, $CuCr_2O_4$, $CuFe_2O_4$, obtained by solid phase interaction between the corresponding oxides, are also efficient catalyst for the thermolysis of AP. Relative efficiency of the catalysts for pure AP decreases in the series: $NiCr_2O_4 > NiFeCrO_4 > NiFe_2O_4$ and $NiFeCrO_4 > NiCr_2O_4 > NiFe_2O_4$ for the mixture.

Al_2O_3, TiO_2 and SnO_2 have no effect on the decomposition of AP, although these oxides catalyze decomposition of perchloric acid. Moreover, there are additives which decelerate the decomposition process of AP. For example, lithium perchlorate or salt of lithium which form equivalent amount of lithium perchlorate by interacting with AP. AP with 1% mass of $LiClO_4$ additive can survive heating at 270 ^{0}C for upto 24 hours without any noticeable indices of thermal decomposition [30]. 10% additive causes strong inhibition.

Effect of various additives on thermal decomposition of AP in the temperature range 170-200 ^{0}C is depicted in Figure 6. Maximum catalytic activity wass shown by nickel and cobalt oxide.

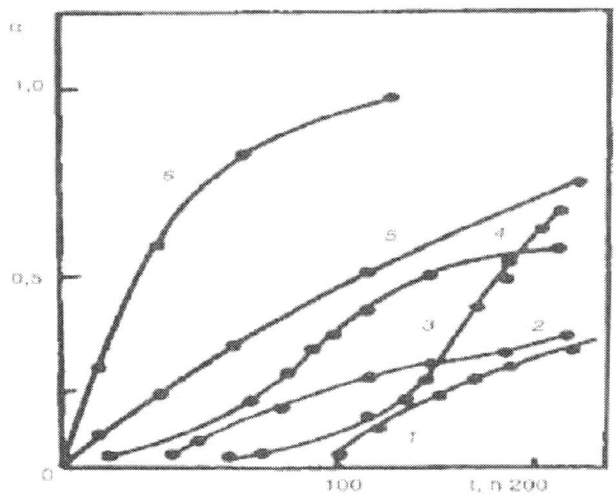

Figure 6. The effect of various oxide additives (30%) on thermal decomposition of AP at 170 ^{0}C; (1) Pure AP, (2) AP + MgO, (3) AP + Cr_2O_3, (4) AP + MnO_2. (5) AP + (Co_2O_3 + Co_3O_4) (6) NiO.

Addition of Cu_2O to AP decreases deflagration delay time due to the catalytic action itself and due to heat evolution during its oxidation to CuO. CuO catalyze the thermolysis of AP even in amounts 0.5% mass and similar to MnO_2; manifest the decomposition of AP till completion. Doping of CuO with Li increases its catalytic activity while doping with Cr_2O_3 decreases (Figure 7). Against this when Cr_2O_3 is doped with CuO, the catalytic activity of Cr_2O_3 increases. Doping of nickel oxide gave same result as in CuO for LTD of AP.

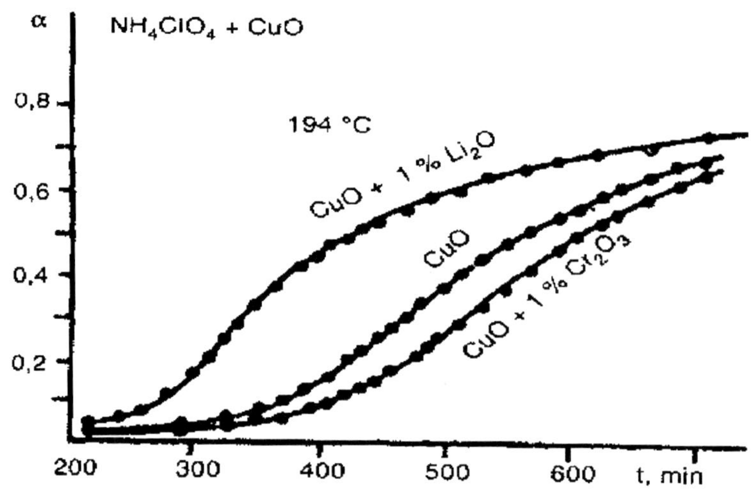

Figure 7. The influence of doping of CuO on its catalytic activity in thermal decomposition of AP at 194 ^0C (molar ratio 1:1).

Ferric oxide and chromium oxide affects mainly not the initial stage but subsequent stages of thermal decomposition, and also affects deflagration delay time.

Oxide catalysts of thermal decomposition are oxidation-reduction catalyst. This is used as an argument to prove the scheme of decomposition mechanism based on the stage of electron transfer from anion to cation.

The additives affect not only the solid AP, but the reactions proceedings in the gas phase. There are grounds to assume that the sub surface reactions, the reactions on the surface and those in the gas phase are closely connected with each other. Gas phase catalization of AP decomposition at 250 ^0C has been proved with NiO additive. The scheme of the experiment is shown in Figure 8. NiO additive is separated with glass filter from AP, yet it catalyze the decomposition. From this figure it is also clear that the larger is the specific surface of the additive, including the outer geometrical surface and the inner surface, the higher is its catalytic action.

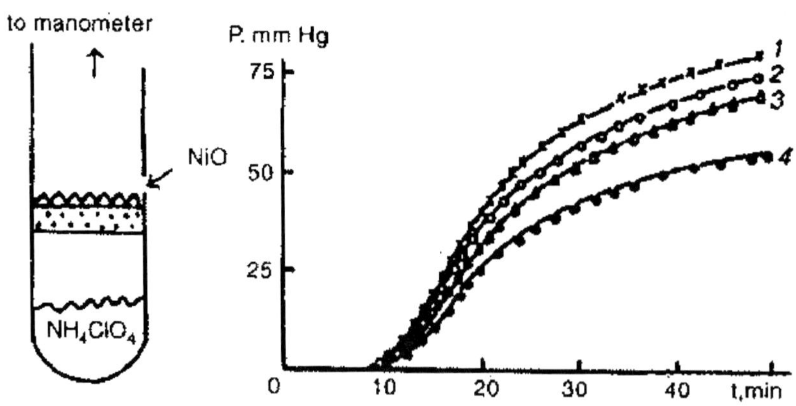

Figure 8. The effect of NiO additive on gas phase thermal decomposition of AP at 250 ^0C. NiO and AP is separated with gap. (1) Pure AP, (2-4) with NiO with the specific surface 2.0, 5.4 and 36.0 m^2/g respectively.

If different weighed amount of NiO and CuO are taken in such a way that the surface of an additive was the same, the effect of an additive on the rate was found to be the same (Figure 9). For example, the action of 1% (with mass) NiO additive with specific surface 36 m^2/g is equivalent to the action of 51% NiO with the specific surface $0.7m^2/g$. Specific surface of catalyst increases, the catalytic activity increases provided the same mass.

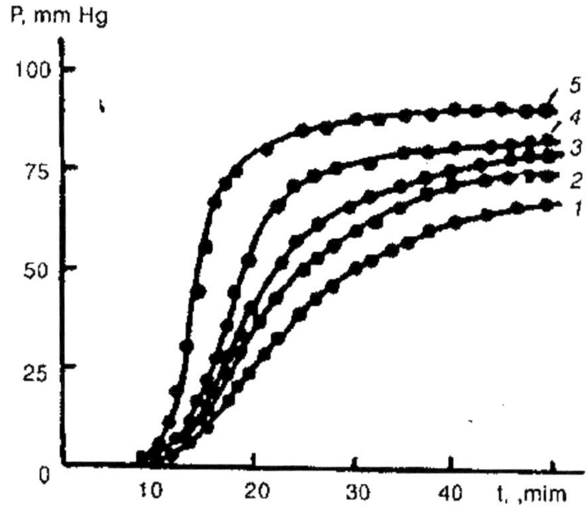

Figure 9. Effect of specific surface of NiO additive on its catalytic activity for thermylysis of AP. (1) Pure AP, (2-5) with NiO having specific surface 0.7, 2.0, 5.4 and 36.0 m^2/g, respectively.

As regards the process carried out in gas phase by oxide additives, there may be either decomposition of $HClO_4$ or oxidation of NH_3 by the products of decomposition of $HClO_4$. The most active catalyst should be considered to be the catalyst accelerating both, the decomposition of the acid and oxidation of ammonia [31].

2.13.2. Effect of Homophase Additives

Additives doped into the lattice of AP during its synthesis or during obtaining crystals, have a substantial effect on physico-chemical properties of AP and may affect its thermal decomposition. Because of this, attempts were made to modify the properties of AP by introducing the ions which could substitute either NH_4^+ or ClO_4^- ion.

Doping of AP with divalent cation of calcium, strontium, zinc and barium accelerates thermal decomposition. Addition of SO_4^{--}, similarly to other di and trivalent anions inhibits the process of thermal decomposition of AP (figure 10).

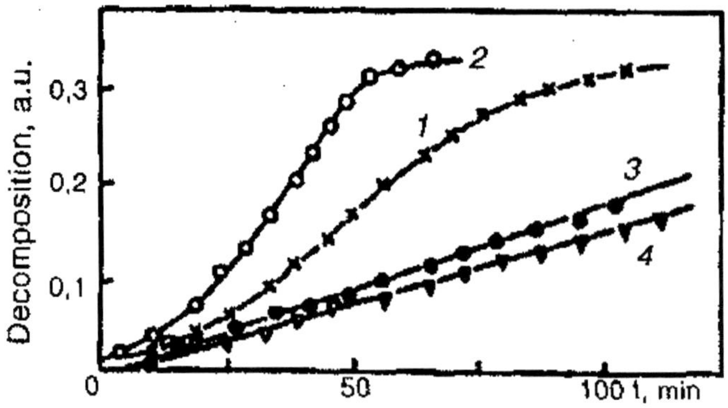

Figure 10. Thermal decomposition of pure and doped AP. (1) Pure AP, (2) AP + 2 × 10^{-2} mol% Ca^{++}, (3) AP + 1 × 10^{-2} mol% SO_4^{--}, (4) AP + 1.5 × 10^{-2} mol% PO_4^{3-}.

The literature available at the present provides evidence that the only mechanism of catalysed AP decomposition is proton transfer mechanism. Reaction starts with proton transfer from cation to anion, followed by migration of proton over the anion sub lattice. Doping of AP lattice with heterovalent cation creates additional number of cation vacancies which should be considered as the creation of additional number of proton traps in the lattice [32]. Since cation vacancy is a site of local negative site, proton should be accumulated in this site forming $HClO_4$. On the contrary, heterovalent anion should eliminate a part of cation vacancies. As a result, proton accumulation at local negative sites will be checked and thermal decomposition will be inhibited. The ions themselves may play roles of donor and acceptor of protons.

If dopant ions are donor or acceptor of protons, they can directly affect the processes participated by protons. Proton donor additives increase the rate of thermal decomposition process. For example, HSO_4^-, $(NH_4)_2SO_4$, $NH_4H_2PO_4$ and $(NH_4)_2HPO_4$. Proton donor additives increase the process rate by accelerating the nuclei formation process (dN/dt) and dose not affect the nuclei formation growth (table 4).

Table 4. The effect of proton additive HSO_4^- on the rate of nuclei formation (dN/dt) and maximal number of nuclei per unit surface area (δmax)

Sample	dN/dt (1/mm^2 min)	δmax (1/mm^2)
AP	45.7 ± 4.0	1003 ± 148
AP + 5 × 10^{-3} mol% HSO_4^-	149.0 ± 10.0	1680 ± 150
AP + 5 × 10^{-2} mol% HSO_4^-	210.0 ± 12.0	2229 ±2 59

Proton acceptor additives decrease the rate of thermal decomposition of AP (figure 11). For example diphenylamine, carbamide and thiocarbamide act as proton acceptors. Inhibiting effect of an anion additive is mainly due to its proton acceptor activity. Generation of anion vacancies during doping is likely to be of secondary importance.

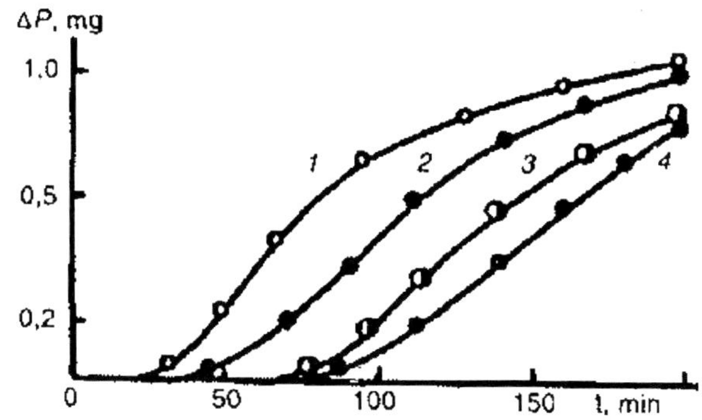

Figure 11. the kinetic curves of AP decomposition at 235 ^{0}C in the presence of organic proton acceptor additives. (1) Pure AP, (2) AP + 1mol% diphenylamine, (3) AP + 1mol% carbamide, (4) AP + 1mol% thiocarbamide.

2.13.2.1 The Effect of Chlorate Ion Additive

The effect of chlorate ion on thermal decomposition of AP is of special importance. First, it is known that the addition of chlorate ion causes a substantial decrease in thermal stability of AP; second, chlorate ion is one of the main products formed during storage and irradiation of AP. Chlorate ion can be introduce into the crystal by means of joint crystallization of ammonium chlorate and perchlorate [33]. Table 5 shows, how is the effect of ClO_3^- on the rate of decomposition. With an increase in 0.1mol% concentration of ClO_3^- ion in AP, the formation of nuclei occurs at such a high rate that it can hardly be recorded. Induction period decreases to 1 min.

Decomposition of chlorate ion occurs much more rapidly in the presence of proton donor additive than without it. In AP, NH_4^+ in every second ion, the route of ClO_3^- decomposition through proton addition is the only one.

Table 5. The effect of chlorate ion additive on the rate of formation of reaction nuclei (dN/dt) and maximal number of nuclei per unit surface area (δ max) at 230^{0}C

Sample	dN/dt (1/mm^2 min)	δ max (1/mm^2)
AP recrystallised five times	21.4 ± 2.1	612 ± 31
AP recrystallised two times	45.7 ± 4.0	1003 ± 142
AP + 8 × 10^{-4} mol% ClO_3^-	146.0 ± 17.0	1860 ± 71
AP + 2 × 10^{-3} mol% ClO_3^-	481.0 ± 52.0	3632 ± 275
AP + 6 × 10^{-2} mol% ClO_3^-	615.0 ± 43.0	4613 ± 280

2.13.3. Effect of Heterophase Precursor of the Catalysts

These are the additives which do not catalyze thermal decomposition but form the catalysts during thermal decomposition of the precursor itself or during its interaction with AP. It was shown [34] (figure 12) that the efficiency of an additive depends on the features of its thermal decomposition. The decomposition is strongly affected by the additives which themselves can easily be decompose to give accelerating thermal decomposition products. For example, $Cu(NO_3)_2$ which form nitrogen oxides. The additives also affect the combustion of the model mix fuel such as AP based propellants. Copper and chromium carbonate and various metal complexes also include this class of additives.

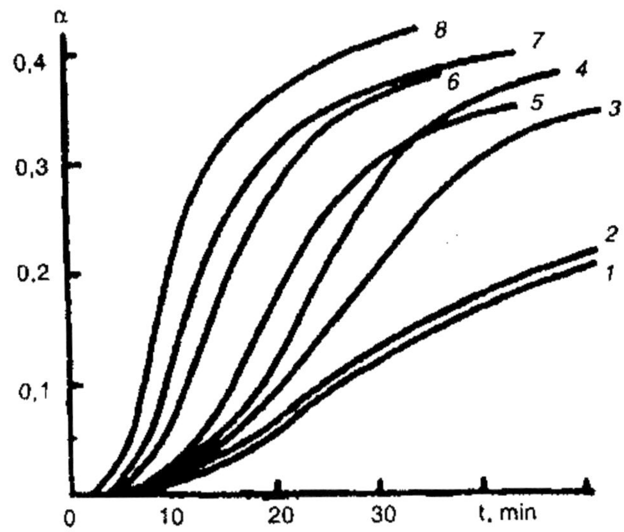

Figure 12. The effect of copper salts as additives (1 mass%) on thermal decomposition of AP at 250^{0}C: (1) copper dichromate and sulphate, (2) pure AP, (3) copper benzoate, (4) copper benzoate heated preliminarily at 250^{0}C, (5) copper lactate, (6) copper oxalate, (7) copper oxide, (8) copper nitrate.

Additives decrease the time of ignition (t_{id}), activation energy and increase the rate of thermal decomposition of AP. Some heterophase additives and their affects on AP decomposition are summarized in tables 6, 7, 8 and 9 [35, 36].

Table 6. Effect of additives on thermal decomposition of AP

Sample	Rate of decomposition (min^{-1}) for mass loss (%)			Catalytic activity (CA)* for mass loss (%)		
	15	20	25	15	20	25
AP	0.57	0.52	0.11	--	--	--
AP + CB	0.57	0.59	0.55	1.00	1.13	5.00
AP + CS	0.57	0.57	0.45	1.00	1.10	4.00
AP + CM	0.55	0.41	0.27	1.00	0.79	2.40
AP + CA	0.70	0.59	0.41	1.22	1.13	3.70

*CA = rate of decomposition of AP or AP + additive / rate of decomposition of AP

CB = copper benzoate; CS = copper salicylate; CM = copper malonate; CA = copper adipate.

Table 7. Variation of ignition delay (t_{id}) with temperature for AP and AP + additives

Technique	Temp (^0C)	Ignition delay (t_{id}) (s)								
		AP	AP + BCC	AP + CuO	AP + CC	AP + Cr_2O_3	AP + CB	AP + CS	AP + CM	AP + CA
Hot plate	303	DNI	70	111	110	DNI	--	--	--	--
	323	DNI	--	--	--	--	45	--	--	70
	329	DNI	49	63	57	DNI	--	--	--	--
	348	DNI	--	--	--	--	28	43	39	55
	354	DNI	36	52	47	DNI	--	--	--	--
	373	DNI	--	--	--	--	25	32	26	38
	378	DNI	25	38	35	--	--	--	--	--
	392	DNI	--	--	--	--	23	24	23	33
	395	DNI	17	27	26	44	--	--	--	--
	427	DNI	16	22	18	33	--	--	--	--
	430	--	--	--	--	--	17	20	19	21
	450	76	11	18	14	26	--	--	--	--
	474	58	10	13	13	20	--	--	--	--
	486	--	--	--	--	--	14	16	14	17
	497	41	08	13	11	16	--	--	--	--
	520	21	08	11	11	16	--	--	--	--
	542	17	07	10	--	--	10	12	10	12
	565	16	--	09	--	14	--	--	--	--
	572	16	--	--	--	--	09	10	10	10
tube Furnace	300	122	97	110	93	--	92	97	120	93
	310	106	92	88	91	146	78	91	104	86
	325	78	64	60	59	141	65	68	74	69
	340	66	60	55	56	105	47	59	53	48
	350	58	56	41	46	77	45	52	51	45

BCC = Basic copper carbonate; CC = Chromium carbonate.

Table 8. Kinetic parameters for AP with and without additives

Sample	ΔT_i*		Activation energy (Ea) (kCal/mol)		Catalytic activity at 255^{0}C (C_A)**
	HP (^{0}C)	TF (^{0}C)	HP Technique	TF Technique	
AP	--	--	19.2	14.1	--
AP + BCC	140	30	8.9	11.5	2.42
AP + CuO	100	30	9.1	13.3	1.77
AP + CC	100	30	9.4	12.3	1.33
AP + Cr_2O_3	40	10	9.5	15.1	0.68
AP + CB	--	--	6.0	10.0	--
AP + CS	--	--	7.0	10.5	--
AP + CM	--	--	8.1	13.1	--
AP + CA	--	--	6.1	12.3	--

* $\Delta T_i = T_i\,(AP) - T_i\,(AP + additive)$;
T_i (AP) = Ignition temperature for pure AP
T_i (AP + additive) = Ignition temperature for catalyzed sample
**C_A = Rate of decomposition of (AP + additive) / Rate of decomposition of AP.

The results of TGA studies on AP and (AP + additives) [39] are given in table 9. The decomposition rate is also found to be increased by the addition of additives (4% by wt.). The AP samples suddenly deflagrate at about 320^{0}C in case of basic copper carbonate and cupric oxide and at about 400^{0}C in case of ammonium dichromate, chromium carbonate and chromium oxide. The increased decomposition of AP with additives may be due to (1) formation of metal perchlorate and/or metal perchlorate amine and (2) exothermic catalyzed decomposition of AP. It has been suggested that proton transfer is the key rate –determining factor during AP decomposition. Metal perchlorate and perchlorate amine are formed between metal oxide and AP and that these accelerate the proton transfer. Furthermore, the decomposition of metal perchlorate amine is exothermic which would enhance AP decomposition. It is obvious that some decomposition must take place of (NH_3 and $HClO_4$) before any of these mechanisms can operate to enhance the rate.

Table 9. TGA of AP + Inhibitors (4% by wt., heating rate 4 ^{0}C/min)

Temp (^{o}C)	Percent decomposed					
	AP	AP + ammonium dichromate	AP + chromium carbonate	AP + chromium oxide	AP + basic copper carbonate	AP + cupric oxide
140	0.76	0.90	0.86	0.80	1.83	1.16
200	3.05	3.18	3.67	3.47	4.21	3.90
240	3.31	4.54	5.42	5.21	5.50	4.68
280	--	5.90	6.95	6.08	7.33	5.46
320	6.87	20.00	14.30	11.52	9.17	7.81
(sudden deflagration)						
360	29.77	30.63	36.47	33.47		
400	44.27	51.18	58.78	57.39		
(sudden deflagration)						
440	96.94					

Table 10 clearly shows that the rate of low temperature thermal decomposition of AP is inhibited to a considerable extent in the presence of $CaCO_3$ and CaO, except for AP plus CaO at 255^{0}C [40]. The decomposition of AP plus CaO is catalysed at 255 ^{0}C up to about 40 min but it is inhibited after that. Recently it has also been reported via dynamic TGA studies that CaO enhances the decomposition of AP up to 280 ^{0}C and decreases thereafter. $CaCO_3$ and CaO inhibit the AP decomposition at 325 and 315^{0}C respectively in the initial stages but at later stages both compounds catalyse the decomposition.

Table 10. Effect of inhibitors (4% by weight) on the thermal decomposition of AP

Temp (^{0}C)	Sample	Rate of decomposition (15%) (min^{-1})	Degree of inhibition (I_A)
255	AP	0.55	--
	AP + $CaCO_3$	0.50	1.1
	AP +CaO	0.68	0.8
275	AP	1.50	--
	AP + $CaCO_3$	0.75	2.0
	AP +CaO	0.55	2.7
300	AP	6.00	--
	AP + $CaCO_3$	0.75	8.0
	AP +CaO	1.25	4.8
325	AP	7.50	--
	AP + $CaCO_3$	1.50	5.0
	AP +CaO	2.50	3.0

$$\text{Rate of decomposition} = \frac{\text{15\% decomposition of the sample}}{\text{Corresponding time for decomposition}}$$

$$I_A = \frac{\text{Rate of decomposition of pure AP}}{\text{Rate of decomposition of AP in presence of inhibitors}}$$

Table 11 clearly shows that the percentage decomposition of PS and AP is lowered with all the inhibitors examined except calcium oxide and barium carbonate [41]. CaO enhances the decomposition of AP and propellants upto 300 ^{0}C and at temperatures higher than 300 ^{0}C, the percentage decomposition is lowered. It has also been observed that the percentage decomposition of AP is enhanced between 360 and 420 ^{0}C when barium carbonate is used as inhibitors.

The inhibition of AP and the propellants decomposition may be due to the formation of metal perchlorates.

$$2NH_4ClO_4 + MCO_3 \rightarrow M(ClO_4)_2 + 2NH_3 + CO_2 + H_2O$$
$$2NH_4ClO_4 + MO \rightarrow M(ClO_4)_2 + 2NH_3 + H_2O$$

where M is calcium, barium and strontium.

It has been observed that NH_3 is evolved even at room temperature when AP is mixed with CaO. The metal perchlorates are more stable to decomposition as compared to AP which is apparent from their heat of formation data. Formation of metal perchlorates is supported by the fact that anhydrous perchlorates of Zn, Cd, Pb, and Ca can be prepared by heating AP with the corresponding oxides or carbonates.

Table 11. TGA of AP + Additives (4% by wt.)

Temp (^0C)	Percent decomposition				
	AP	AP + $CaCO_3$	AP + $BaCO_3$	AP + $SrCO_3$	AP + CaO
140	0.35	0.35	0.36	0.36	0.35
160	0.71	0.69	--	--	2.85
180	1.07	--	0.73	1.09	4.28
200	2.14	1.75	1.10	--	5.35
220	2.50	2.10	--	1.46	7.14
240	2.85	--	1.47	2.56	--
260	3.21	2.45	2.94	2.93	7.50
280	3.57	--	3.67	3.29	8.57
300	3.92	3.15	4.41	4.39	10.0
320	14.64	11.22	5.88	8.05	12.50
340	24.21	16.49	15.07	19.41	21.42
360	28.92	18.58	36.02	21.24	--
380	30.71	20.00	41.17	23.07	22.85
400	32.50	21.75	44.85	26.73	27.14
420	39.64	27.01	54.04	33.33	32.14
440	56.07	*	*	51.81	35.71
460	*	--	--	58.34	39.28
480	--	--	--	*	*

The kinetic parameters reported in table 12 [39] show that rate constants for AP plus $CaCO_3$ and AP plus CaO decrease when compare with pure AP in acceleratory and deceleratory periods at 255, 275 and 300^{0}C.

Table 12. Kinetic parameters for thermal decomposition of AP, AP + $CaCO_3$ and AP + CaO

Temp (^oC)	Sample	Rate constant (min^{-1})	
		$k_a \times 10^3$	$k_d \times 10^3$
255	AP	4.7	0.30
	AP + $CaCO_3$	3.2	0.20
	AP +CaO	3.0	0.28
275	AP	6.7	0.30
	AP + $CaCO_3$	5.0	0.11
	AP +CaO	2.5	0.21
300	AP	20.0	0.50
	AP + $CaCO_3$	8.0	0.37
	AP +CaO	4.5	0.36

2.13.4. Effect of Homophase Precursor of Catalyst

These are the additives introduced into the lattice of AP which itself does not catalyze the thermal decomposition of AP, but when subjected to heat they decompose to give their own decomposition product which catalyze the thermolysis of AP. This class of additives includes Ag^{+} ions, $KMnO_4$, K_2MnO_4, K_2CrO_4, $K_2Cr_2O_7$, KIO_4, KIO_3, perchlorate of cobalt, nickel and iron etc. Silver ions are transformed into metal silver during thermal decomposition. MnO_4^{-}, MnO_4^{--}, CrO_4^{--}, $Cr_2O_7^{--}$ etc form oxide catalyst as a result of their decomposition. MnO_4^{-} has got special attention because it is isomorphous to ClO_4^{-} ion, so that it easily enters into the lattice of AP and it is thermally less stable than perchlorate and decompose to results MnO_2, which is strong catalyst for AP decomposition. It has been shown that isomorphically introduced $KMnO_4$ has a strong effect on the thermal decomposition than mechanically mixed. The highest effect is caused when 50% of ClO_4^{-} is substituted by MnO_4^{-}.

At initial stage, $Cr_2O_7^{--}$ ion does not decompose [37, 38] and acts as active proton acceptor and inhibits the formation of reaction nuclei. But, at higher temperatures, dichromate ion decompose to give Cr(III) oxide which catalyze the thermolysis of AP.

Three possible ways of the influence of additives were discussed so far:

(a) Formation of cation vacancies during the doping with doubly charge ions into lattice of AP.

(b) Perchlorates of transition metals decompose to form oxides. These oxides are catalysts of low temperature thermal decomposition of AP. This is manifested by the similarity that thermal decomposition curve of AP with transition metal perchlorate additives and with transition metal oxide additives (figure 13)

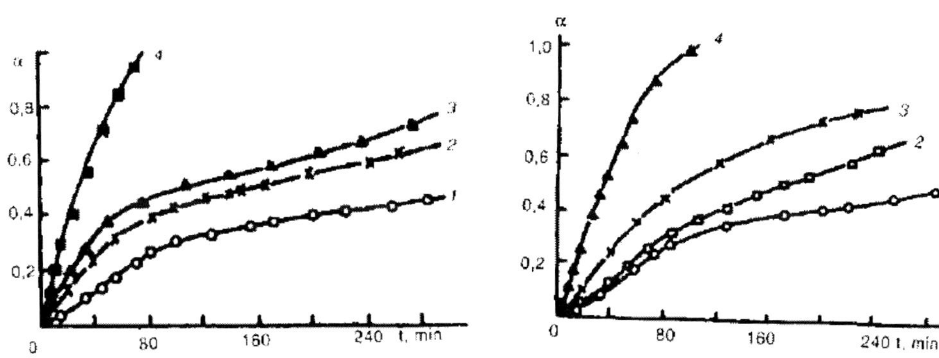

a. b.

Figure 13. Thermal decomposition of AP at 275^{0}C (a) doped with; (1) pure AP, (2) Fe^{++}, (3) Ni^{++}, (4) Co^{++} ion (b) with oxide additive as heterophase; (1) Pure AP, (2) 1% Fe_2O_3, (3) 1%Ni_2O_3, (4)1%Co_2O_3.

2.14. THE INFLUENCE OF CRYSTAL SIZE AND MECHANICAL TREATMENT ON THERMAL DECOMPOSITIOM OF AP

A decrease in the size of the crystallites the rate of thermal decomposition increases but after a certain size it decreases [19]. A reason for such a specific dependence is likely to be fact that the process of thermal decomposition of AP starts not on the surface but in the sub surface layer on potential reaction centres. In order to put these reaction centres into operation, it is necessary that primary decomposition products accumulate in it. But crystal dispersing brings about two opposite factors:

(i) Removal of reaction products occurs through the surface. So a decrease in crystal size improves the reaction conditions.

(ii) When crystallite size becomes of the same order of magnitude as the distance from the potential centres of the surface, the formation of new dislocations as a result of the accumulation of reaction products become worse, causing a decrease in reaction rate.

Inspite of these factors, density of dislocations in AP crystals changes as a result of mechanical treatment. Mechanical treatment of AP with shock wave causes acceleration of the LTD, as the exo peak in DTA curve of mechanically treated AP lies at 240 ^{0}C while for not treated mechanically this peak is shifted to above phase transition temperature [37]. Amorphisation of the crystals under mechanical treatment leads to a decrease in the starting decomposition temperature [40]. Recompression of AP modify the concentration of potential centres of reaction which significantly modify the rate of thermal decomposition of AP [41].

2.14. CONCLUSIONS

AP exists as two modifications: orthorhombic below 240 ^{0}C and cubic above 240^{0}C. Experiments show that the mechanism of conductivity in AP is proton type. Primary stage of the process of thermal decomposition of AP is proton transfer as NH_3 and $HClO_4$ were detected as primary products of decomposition. Due to radiolysis, ClO_3^- ion accumulate in the crystal of AP. ClO_3^- ions are one of the most efficient traps for proton regenerating during thermolysis. So irradiations affect thermolysis of AP. Homophase or heterophase additives also affect the thermal decomposition. Additives may decrease the time of ignition, activation energy and increase the rate of thermolysis of AP. A decrease in the size of crystallites, the rate of thermal decomposition increases but after a certain size it decrease.

REFERENCES

[1] M. W. Evans, R. B. Beyer and L. Mc Culley, *J. Chem. Phys.*, 40 (9) (1964) 2431.

[2] J. A. Ibers, *J. Chem. Phys.*, 32 (1960) 1448.

[3] A. V. Dubovitsky, N. Ya. Buden and G. B. Manelis, russ. *J. Struct. Chem.*, 5 (1964) 40.

[4] V. Boldyrev, *Thermochim. Acta,*100 (1986) 330.

[5] J. N. Maycock andV. R. Pai Verneker and G. A. Gorzinsky, *Solid State Commun.*, 5 (4) (1967) 225.

[6] J. N. Maycock and V. R. Pai Verneker, *Proc. Roy. Soc.,* A307 (1968) 303.
[7] P. Zirkind and E. S. Freeman, *Nature,* 199 (1963) 1280.
[8] H. Wise, *J. Phys. Chem.,* 71 (1967)2843.
[9] V. V. Boldyrev and E. F. Khairetdinov, *J. Inorg. Nucl. Chem.,* 31 (1969) 3332.
[10] E. F. Khairetdinov, A. A. Medvinsky and V. V. Boldyrev, *Kinetica I Kataliz.,* 11 (5) (1970) 1343.
[11] E. F. Khairetdinov, V. V. Boldyrev and A. I. Burshtein, *J. Solid State Chem.,* 10 (1974) 288.
[12] G. Singh, I. P. S. Kapoor, J. Singh and J. Kaur, *Indian J. Chem.,* B (2000) 1.
[13] M. Dode, B*ull. Soc. Chim. France,* 5 (1938)170.
[14] M. Dode, B*ull. Soc. Chim. France,* 5 (1938(176.
[15] G. L. Pellet and W. R. Cofer, seventh AIAA Aerospace Science Meeting, New York, 20 January, (1969) 141.
[16] A. V. Raevsky, A Collection of papers, G. B. Manelis (Ed.), Chernogolovka: Institute of Chemical Physics AS USSR, (1981) 30.
[17] A. K. Galway and M. A. Mohamed, Proc. *Roy. Soc. Lond.,* A396 (1984) 425.
[18] A. K. Galwey and P. W. M. Jacobs, *Trans. Faraday Soc.* 55 (1959) 1165.
[19] L. Bircomshaw and B. Newman, *Proc. Roy. Soc.,* A227 (1955) 228.
[20] P.W.M. Jacobs and H. M. Whitehead, *Chem. Rev.,* 69 (1969) 551.
[21] E. Freeman, D. Anderson and J. J. Camposi, *J. Phys. Chem.,* 64 (1960)1727.
[22] E. Freeman and D. Anderson, *J. Phys. Chem.,* 65 (1961) 1662.
[23] A. I. Skorik, V. V. Boldyrev, V. F. Komarov and A. I. Baranov, Izv. VUZov. Ser. *Khimiya I Khimicheskaya Tekhnologiya.* (7) (1968) 744.
[24] A. I. Skorik, V. V. Boldyrev and V. F. Komarov, *Kinetica I kataliz.* 8 (B. 6) (1967) 1258.
[25] P. Herley, P. W. Levy, in Proceedings of the 7th International Symposium on the reactivity of Solids, Bristol, (1972) pp. 387.
[26] K. C. Patil, V. R. Pai verneker and S. R. Jain, *Indian J. Chem. Soc.,* A 16A (2) (1978) 109.
[27] F. Solymosi and M. Rauics, *Combust. Flame,* 10 (1966) 398.
[28] P. Herley, P. W. Levy, Quantitative study on radiation induced dislocations and the decomposition kinetics of ammonium perchlorate, in: Proceedings of the 7th International Symposium on the Reactivity of Solids, Bristol, 1972, pp 387-398.
[29] F. solymosi, Akademiai Kiado, Budapest, (1977) 195.
[30] T. V. Mulina, *Thesis, Cand. Sci. Chem.* Novosibirsk, (1976) 80 (In Russian).
[31] V. I. Plyusnin, A. V. Boldyrebva and V. V. Boldyrev, *Kinetika I kataliz.,* 10 (4) (1969) 761.
[32] G. Singh, R. R. Singh, A. P. Rai and I. P. S. Kapoor, *J. Therm. Anal.* 36 (1990) 2539.
[33] Gurdip Singh, S. K. Vasudeva and I. P. S. Kapoor, *Indian J. Technol.* 29 (1991) 589.
[34] G. Singh, I.P. S. Kapoor, J. Singh and J. Kaur, *Indian J. Chem.,* B (2000) 1.
[35] V. Boldyrev, *Thermochim. Acta,* 100 (1) (1986) 330.
[36] A. V. Raevsky, G. B. Manelis, V. V. Boldyrev and A. Votinova, Dokl. AN SSSR 160 (5) (1965) 1136.
[37] S. M. Reshetnikov, S. G. Voronchikhin and T. D. Korotkova, kazan (2) (1979) 53 (In Russian).
[38] G. Pramond and M. P. Kannan, *J. Indian Chem. Soc.* 77 (5) (2000) 261.

[39] R. P. Rastogi, Gurdip Singh and Ram Raj Singh, *Combust. Flame,* 30 (1977) 117.
[40] R. P. Rastogi, Gurdip Singh and Ram Raj Singh, *J. Appl. Chem. Biotechnol.* 28 (1978) 889.
[41] R. P. Rastogi, Gurdip Singh and Ram Raj Singh, *Indian j. Chem.,* 15A (1977) 845.

Chapter 3

CHEMISTRY AND THERMOLYSIS OF SALTS OF PERCHLORIC ACID

ABSTRACT

This chapter deals with the thermal decomposition of nitronium, nitrosonium, hydrazinium, guanidinium, nitroguanidinium, alkyl and aryl substituted ammonium perchlorates.

3.1. INTRODUCTION

Perchlorate ion (ClO_4^-) is a strong plosophoric group. Many of the perchlorates (salts of perchloric acid) find technological applications in explosives, pyrotechnics and propellants. Ammonium salts of perchlorates are perhaps the most importnat among perclorates in terms of energetic material applications. This is because of the combination of ammonium acting as fuel and perchlorate group playing as a perfect oxidiser; two priori for an energetic material. Ammonium perchlorate is the most widely employed perchlorate and its thermolysis is described in chapter 2. Thermolysis of some of the other important perchlorates are described in this chapter.

3.2. NITRYL OR NITRONIUM PERCHLORATE

Nitronium perchlorate is the simplest inorganic perchlorate, has the highest oxygen balance (OB) and is very suitable oxidizer for propellants. Initially, it was prepared by reacting nitric acid and perchloric acid. Gordon and Spinks obtained it by mixing ClO_2 with O_3 and NO_2 [1]. Goddard *et al* prepared by treating dry N_2O_5 with anhydrous $HClO_4$ at -80 ^0C [2].

The mechanism of thermal decomposition of nitronium perchlorate was initially investigated by Gordes [3] and then by Marshall and Lewis [4]. It has been reported that low temperature decomposition goes via nitrosonium perchlorate ($NOClO_4$) formation. Following are the decomposition reactions proposed by Maycock and Verneker [5]:

$NO_2ClO_4 \rightarrow NOClO_4 + 1/2O_2$
$2NOClO_4 \rightarrow NO_2ClO_4 + ClO_2 + NO_2$
$NO_2ClO_4 \quad NO_2 + O_2 + ClO_2$

The kinetic analysis of TG data showed the involvement of branching- nuclei mechanism and ion pair evaporation [5]. Nitronium perchlorate reacts violently with many organic compounds causing ignition or explosion. When it is dissolved in nitromethane, it can be used as a nitrating agent.

3.3. Nitrosyl or Nitrosonium Perchlorate

Nitrosonium perchlorate is a strong endothermic compound and was prepared by passing a mixture of NO and NO_2 into 72% $HClO_4$. The LTD and HTD of nitrosonium perchlorate proceed as follows:

at $<100\ ^0C$, $2NOClO_4 \rightarrow 2ClO_2 + N_2O_5 + 1/2O_2$
at $>100\ ^0C$, $2NOClO_4 \rightarrow Cl_2 + N_2O_4 + 3O_2$

The overall decomposition was given by Glasner et al. [6] as:

$4NOClO_4 \rightarrow 4NO_2 + 3ClO_2 + 1/2Cl_2 + 3O_2$

$NOClO_4$ undergoes ignition/explosion with ethanol, acetone, dry ether and primary amines.

3.4. Hydrazinium Mono- and Diperchlorates

Hydrazinium monoperchlorate (HP-1) is a crystalline solid, has a monoclinic structure. Levy *et al.* prepared HP-1 by titrating 85% hydrazine hydrate with 48% $HClO_4$ until a pH of 3.2 was reached and then precipitated it by pouring into 2-propanol at 0 ^0C [7]. Thermal decomposition of Hp-1 involves proton transfer mechanism [8]. The decomposition of $HClO_4$ was assumed to be the primary step in the thermolysis and the overall decomposition of HP-1 is as follow[9]:

$2N_2H_5ClO_4 \rightarrow 0.8NH_4ClO_4 + 0.7O_2 + 1.6N_2 + 0.6Cl_2 + 3.4H_2O$

Later Verneker *et a*l. [10, 11] have proposed the following reaction:

$5N_2H_5ClO_4 \rightarrow 4NH_4ClO_4 + HCl + 3N_2 + 4H_2O$

HP-1 has been found to be more sensitive to shock and less sensitive to impact.

Hydrazinium diperchlorate (HP-2) was prepared by adding 72% $HClO_4$ to 85% hydrazine hydrate at 5 ^0C[12]. The peculiar feature of the decomposition of HP-2 is the extremely long induction period (77 hours at 120 ^{0}C) after which the reaction suddenly accelerates and the compound completely decomposes within few seconds as follows:

$$12N_2H_6(ClO_4)_2 \rightarrow 4NH_4ClO_4 + 12HClO_4 + 22H_2O + 10N_2 + 5O_2 + 4Cl_2$$

Decomposition of $HClO_4$ was postulated as rate determining step [13]. The LTD and HTD of HP-2 can be represented [14] as:

$$N_2H_6(ClO_4)_2 \rightarrow N_2H_5ClO_4 + HClO_4\ (<140\ ^0C,\ LTD)$$

$$24N_2H_6(ClO_4)_2 \rightarrow 8NH_4ClO_4 + 26HClO_4 + 42H_2O + 6N_2 + 14N_2O + 7Cl_2 + H_2\ (>140\ ^0C,\ HTD)$$

3.5. Guanidinium and Nitroguanidinium Perchlorates

Guanidinium perchlorate (GP) is a relatively stable and non-hygroscopic energetic perclorate having moderate sensitivity to shock and is capable of producing violent explosion. Crystallographic modification of GP occurs at 180 ^{0}C from orthorhombic to cubic. GP can be prepared by fusing dicyanamide with AP at 150-160 ^{0}C. Glasner and Makovky reported that HCl, CO, CO_2, N_2, NH_3, O_2 and H_2O are the major products of decomposition of GP [15].

$$2CH_5N_3HClO_4 \rightarrow 2HCl + 5H_2O + CO + CO_2 + 3N_2 \ \ldots.(1)$$

$$2CH_5N_3HClO_4 \rightarrow 2HCl + 2NH_3\ 2CO_2 + O_2 + 2N_2 + 2H_2O \ \ldots.(2)$$

Generally, latter reaction predominates with increasing temperature. Udupa has confirmed these reactions by TG, DTA, mass spectrometry and X-ray diffractometry [16].

Nitroguanidinium perchlorate (NGP) was prepared from a mixture of nitroguanidine and $HClO_4$ [17]. The thermal decomposition of NGP begins in the solid phase and proceeds as follows:

```
        NH–NO2
        |                →  CH5N3.HClO4 + CO2 + NO2 + 3/2N2
2 H2N–C=NH.HClO4
```

NGP has been reported to be more unstable as compared to GP.

3.6. Alkyl Substituted Ammonium Perchlorates

The preparation of a large number of substituted ammonium perchlorates have been described in the literature. Some of these have been used in solid propellants as oxidizers. These compounds sublimate and deflagrate at lower twmperatures, and at higher temperatures they readily explode. Some of them even explode on slight shock or friction. These compounds are prepared by mixing $HClO_4$ and the appropriate amine. The thermal properties of some of them are discussed below:

3.6.1. Methyl Ammonium Perchlorates

It has been reported that substitution of methyl group into NH_4^+ cation stabilizes the perchlorate salts thermally in the following order as compared to ammonium perchlorate[20]:

$$NH_4ClO_4 < CH_3NH_3ClO_4 < (CH_3)_2NH_2ClO_4 < (CH_3)_3NHClO_4$$

Among these substituted methyl ammonium perchlorates, monomethyl ammonium perchlorate is the most sensitive to impact and friction. These compounds decompose in to the respective methyl substituted amines and $HClO_4$ by proton transfer.

$$CH_3)_nNH_{4-n}ClO_4 \xrightarrow{\varepsilon} (CH_3)_nNH_{3-n} + HClO_4 \xrightarrow{\acute{\varepsilon}} Products$$

where n is the number of methyl groups (1-3), ε is the dissociation energy and έ is the net dissociation energy. Thermal stabilities of these salts increase with the increase in the number of methyl groups. During thermolysis of methyl ammonium perchlorates, in addition to proton transfer, occurrence of methyl group transfer has also been reported.

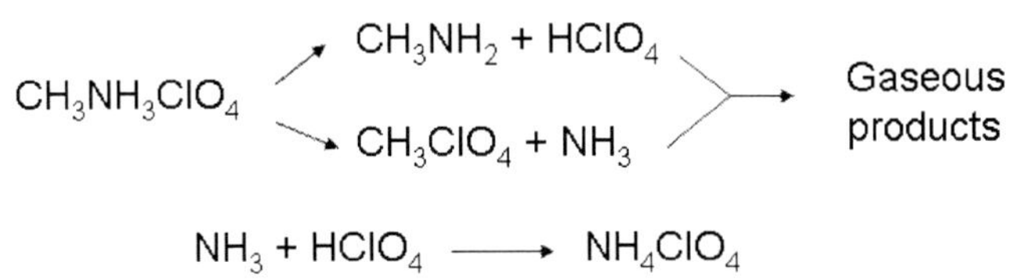

Methyl perchlorate, being unstable in this temperature range, may decompose as soon as it is formed. In each case, there is an additional mode of decomposition giving rise to small contributions of HCl, NO_2, H_2O, O_2, N_2 and N_2O.

3.6.2. Miscellaneous Alkylamine Perchlorates

Amine perchlorates have both the combustible part (fuel part) and the oxidizer part in the same molecule. These are generally much more violent in their explosive behaviour. Most of the perchlorates decompose at lower temperatures and explode suddenly when subjected to higher temperatures. Tetraethyl ammonium perchlorate (TEAP) undergoes phase transition at 98 ^{0}C and explodes at 298 ^{0}C and fragmentation occurs through three routes namely (i) perchlorate anion and its disintegration products, (ii) tetraethylammonium cation and its products involving dealkylation process and (iii) the oxidation of organic group by the perchlorate.

The decomposition of ethylenediammonium perchlorate (EDP) involves four stages[20]. Proton transfer takes place in the temperature range of 275-310 ^{0}C and 320-360 ^{0}C and the whole material is oxidized to gaseous products. It has been reported that thermal decomposition of triethyl ammonim perchlorate and diphenyl ammonium perchlorate under go by proton transfer mechanism. Pentaerythrityl tetrammonium perchlorate undergoes decomposition in a complicated fashion [20]. The decomposition products such as HCl, CO_2, ClO_3 and H_2O were detected but the nitrogen containing products were not traceable which suggests that N_2 may be the product and/or it is retained as a part of condensed phase reactions. Propyl-1,3-diammonium perchlorate (PDP) undergoes a solid-solid phase transition at 150 ^{0}C and significantly changes the lattice structure and the ion dynamics. The decomposition of PDP starts immediately after melting at 280 ^{0}C forming HCl, CO, CO_2 and H_2O as thermodynamically stable products.

3.6.3. Anilinium Perchlorates

Jain *et al.* have studied the thermal decomposition and explosion characteristics of ring (mono) substituted perchlorate (RSAP) by DTA, explosion delay and impact sensitivity measurement [21]. The decomposition temperature and activation energy for explosion have been found to increase with increasing basicity of the corresponding arylamine. The involvement of proton transfer from anilinium ion to perchlorate ion was reported during decomposition and explosion of these salts. Very recently, Singh *et al.* have prepared and characterized a large number of RSAP [22, 23]. The thermolysis of these salts was undertaken by thermoanalytical techniques, explosion characteristics (by explosion delay), impact and friction sensitivity measurements. The explosion temperatures, energies of activation for decomposition and impact sensitivity data ($h_{50\%}$) were found to be linearly related to pKa values of the corresponding arylamines and substituent constants, σ^+ and σ^-. Proton transfer process has been proposed as the rate determining step. It seems that oxidation reaction between arylamines and $HClO_4$ and /or its decomposition products cause explosion. A reaction Scheme 1 has been proposed based on the formation of reaction intermediates during decomposition and explosion.

Dimethylanilinium perchlorates (DMAP) have also been prepared and characterized [24]. Thermolysis of these salts have been investigated using TG, DTA, explosion delay, explosion temperatures, impact and friction techniques. It seems that thermolysis of DMAP involves competitive decomposition reaction paths. The proton transfer seems to control the decomposition induced by impact and free radical mechanism predominantly involved in thermal reactions induced by heat as shown in Scheme 2.

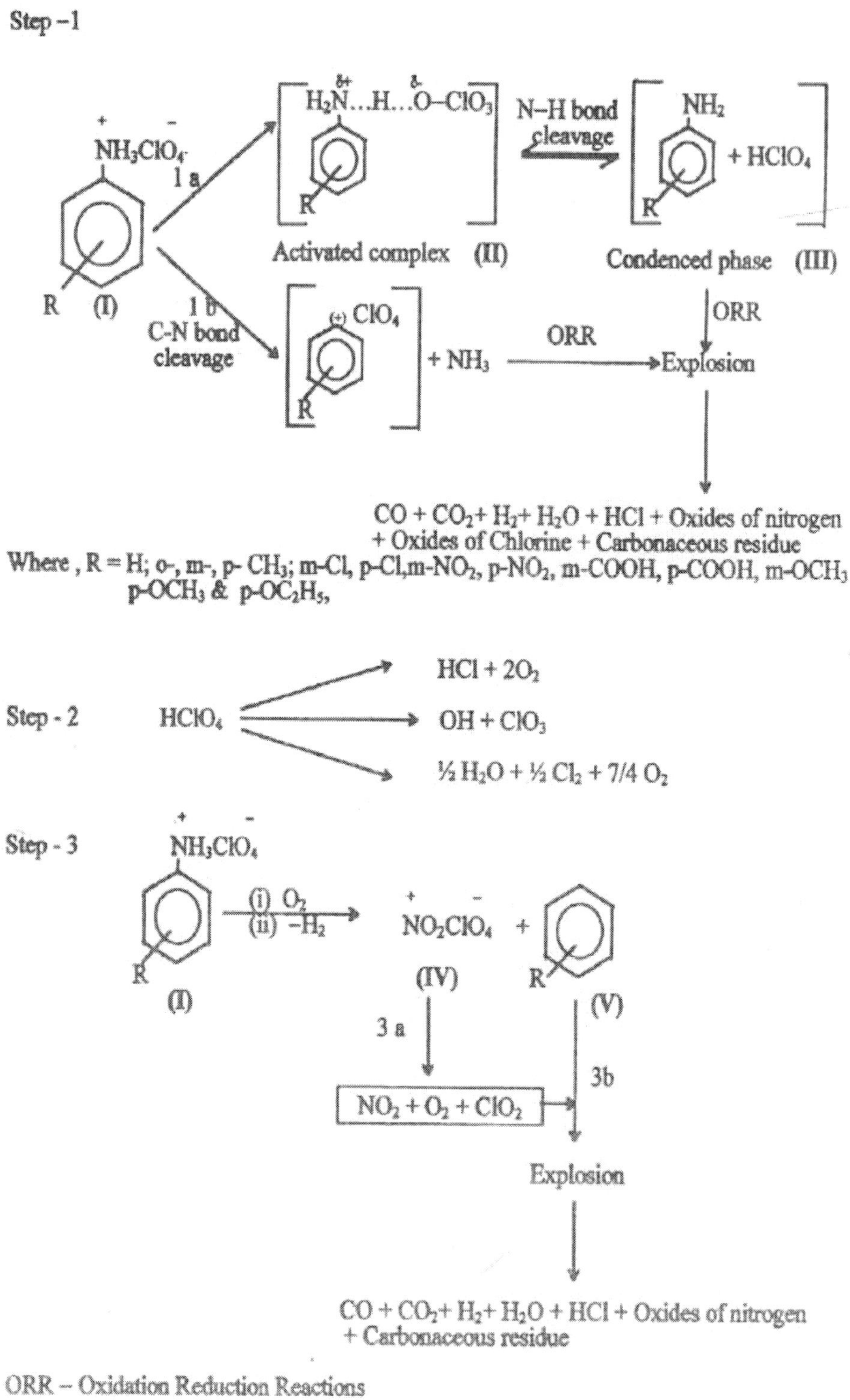

Scheme 1. Schematic representation of the thermal decomposition pathways of ring substituted arylammonium perchlorates.

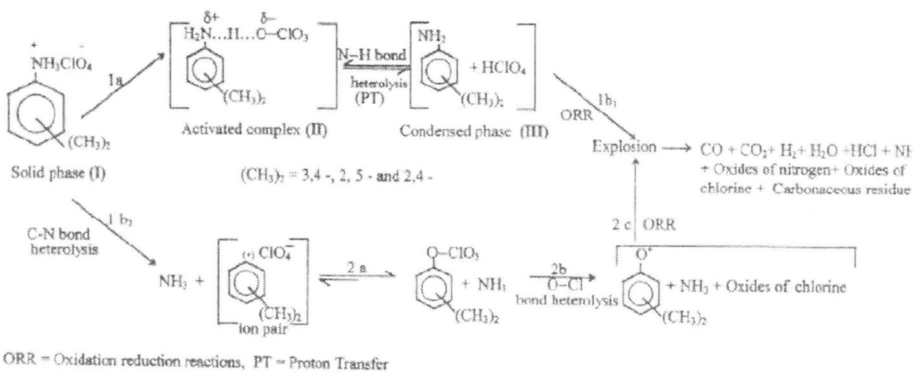

Scheme 2. Schematic representation of the thermal decomposition pathways of dimethylanilinium perchlorates.

3.7. Conclusions

Low temperature decomposition of nitronium perchlorate proceeds via nitrosonium perchlorate. Nitrosonium perchlorate decompose to give NO_2, ClO_2, Cl_2 and O_2. Hydrazinium mono and di-perchlorate initially decomposes to give ammonium perchlorate which further decomposes to various decomposition products. Guanidinium perchlorate undergo transformation at 180 ^{0}C from orthorhombic to cubic. Its major decomposition products are reported to be HCl, CO, CO_2, N_2, NH_3, O_2 and H_2O. Thermal stabilities of methyl ammonium salts increases with the increase in the number of methyl group. During thermolysis of these, in addition to proton transfer occurrence of methyl group transfer has also been reported. Proton transfer process has been proposed as rate determining step in the thermolysis of anilinium and dianilinium perchlorates.

References

[1] W. E. Gorden and J. W. T. Spinks, *Can. J. res.,* b18 (1940) 358.

[2] D. R. Goddard, E. D. Hughes and C. K. Ingold, *J. Chem. Soc.,* (1950) 2559.

[3] H. F. Cordes, *J. Phys. Chem.,* 67 (1963) 1993.

[4] M. D. marshall and L. L. Lewis, *Advances in Chemistry Series,* No. 64, am. Chem. Soc., Washington, (1968) 82.

[5] J. N. Maycock and V. R. Pai Verneker, *J. Phys. Chem,* 72 (1968) 4004.

[6] A. Glasner, J. Pelly and M. Steinberg, *J. Inorg. Nucl. Chem.,* 31 (1969) 3395.

[7] J. Levy, g. Vol Elbe, R. Friedman, T. Wallim and S. *J. Adams, Advan. Chem. Ser.,* 54 (1966) 55.

[8] A. Hermoni and A. Salmon, *Israel J. Chem.,* 2 (1964) 264.

[9] P. W. M. Jacobs and A. Russel-Jones, *Canad. J. Chem.,* 44 (1966) 2435.

[10] V. R. Pai Verneker, A. N. Sharma and S. R. Jain, *Therchim. Acta.,* 17 (1976) 319.

[11] V. R. Pai Verneker, A. N. Sharma and S. R. Jain, *J. Therm. Anal.,* 11 (1977) 177.

[12] E. T. Mcttale, S. J. Adams, G. Elbe and J. B. Levy, *Combust. Flame,* 11 (1967) 141.

[13] A. A. Zinovjer and A. B. Tesentisper, Russ. *J. Inorg. Chem.*, 4 (1959) 329.
[14] V. R. Pai Verneker, A. N. Sharma and S. R. Jain, *propel. Explos.*, 1 (1976) 107.
[15] A. Glasner and A. Makovky, *Bull. Res. Council Israel,* 3 (1953) 89.
[16] M. R. Udupa, *propel. Explos. Pyrotech.,* 8 (1983) 109.
[17] V. ya Rosolovskii and K. V. Titora, Russ. *J. Inorg. Chem.* 12 (1967) 675.
[18] H. Burton and P. F. G. Praill, *Analyst,* 80 (1955) 4.
[19] J. C. Schumacner, Reinhold, New York, (1960).
[20] Gurdip Singh, Inder Pal Singh Kapoor, Jyotsna Singh and Jaspreet kaur, Indian *J. Chem.* 39B (2000) 1.
[21] S. R. jain, M. V. Rao and V. R. Pai Verneker, *Combust. Flame,* 35 91979) 289.
[22] G. singh, I. P. S. Kapoor, *J. Phys. Chem..* 96 (1992) 1215.
[23] G. singh, I. P. S. Kapoor and S. M. Mannan, *J. Therm. Anal.,* 46 (1996) 1751.
[24] G. singh, I. P. S. Kapoor and S. M. Mannan, *J. Energ. Mater.* 13 (1995) 141.

Chapter 4

CHEMISTRY AND THERMOLYSIS OF SALTS OF NITRIC ACID

ABSTRACT

The nitrates (salts of nitric acid) demonstrate explosive properties and can act as readily available oxygen carriers. The chemistry and thermolysis of some of the nitrate salts are described below.

4.1. INTRODUCTION

Ammoium nitrate (AN) is one of the most commercially important ammonium compounds in terms of usage. It finds extensive use in the area of nitrogen fertilizers and explosives. Its principal use in propellants is restricted to low burning rate, low performance applications, such as gas generator for turbo pumps of liquid propellant rocket engines or emergency starters for jet aircraft [1]. Most solid motors currently in use employ ammonium perchlorate (AP) as the main oxidizer. Of late, with the increasing concern for eco-friendly, chlorine free propellants, there is a growing interest in AN, although it is well known that inorganic nitrates are low performance oxidizer as compared to perchlorates and produces innocuous products of combustion. AN is hygroscopic, undergoes a room temperature phase transition involving a significant volume change and burns very slowly. However, most of these problems of AN is surmountable and hence there is renewed interest in this compound [2, 3].

4.2. PROPERTIES OF AMMONIUM NITRATE

Some of the physical properties and performance parameters of AN, which is relevant to its application as a HEM is summarized in table 1.

Table 1. Properties of ammonium nitrate

Property	Value
Molecular formula	NH_4NO_3
Molecular weight	80
Heat of combustion	346 cal g^{-1}
Heat of formation	1098 cal g^{-1}
Heat of explosion	346 cal g^{-1}
Heat of fusion	18.23 cal g^{-1}
Density	1.725 g cm^{-3}
Melting point	169.6 ^{0}C
Specific volume	0.580 cm^3 g^{-1}
Solubility in water at 20 ^{0C}	66 g / 100 g
Oxygen content	60%
Available oxygen	20%
Estimated flame temperature	1500 ^{0}C
Detonation velocity	1250-4650 m s^{-1}
Coefficient of thermal expansion at 20 ^{0}C	9.82 X 10^{-4} % / ^{0}C

A typical TG-DSC curve of AN is given in Figure 1

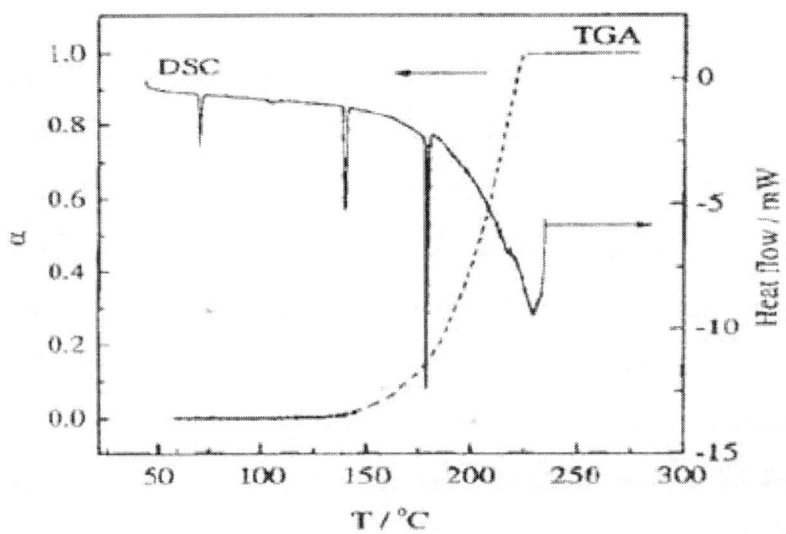

Figure 1. Typical TG-DSC thermogram of AN under inert atmosphere.

Under ordinary pressure, AN has not less than five stable polymorphic transformations [4-8] as shown below in Scheme 1.

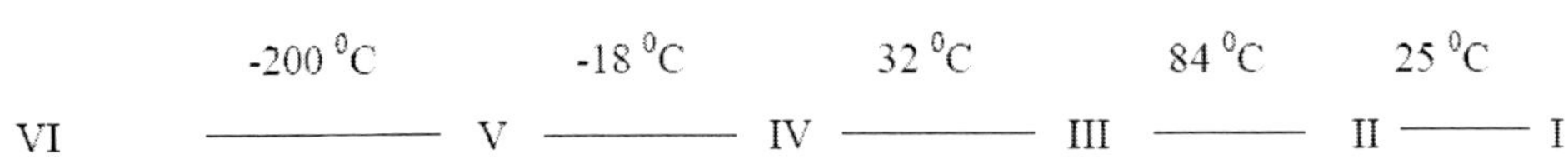

Scheme 1. Phase transformations in AN.

After the initial polymorphic transitions, AN melts at 169.5 ^{0}C [9]. Decomposition of AN takes place above 200 ^{0}C.

The mechanism of thermolysis is highly dependant on the experimental conditions. However, most mechanism proposed for AN decomposition assume the production of ammonia and nitric acid in the primary step and subsequent oxidation of NH_3 by the decomposition products of HNO_3.

AN decomposes upto 98% inbetween 210-260 ^{0}C into nitrous oxide and water [10] This decomposition often takes place explosively above 300 ^{0}C and the products are also different. According to Shah and Oza [11], the following reactions take place:

(i) Dissociation of AN to ammonia and nitric acid
(ii) Production of nitrogen
(iii) Decomposition into nitrous oxide and water.

Feick and Hainer suggested the following equations for thermal decomposition of AN [12]:

at 170 ^{0}C, $NH_4NO_3 \leftrightarrow NH_3 + HNO_3$ $\Delta H = 35.4$ kcal/mol
at 290 ^{0}C, $NH_4NO_3 \leftrightarrow N_2O + 2H_2O$ $\Delta H = 16.4$ kcal/mol

The thermal decomposition of AN in the temperature range 40-200 ^{0}C under reduced pressure has been found to be strongly temperature dependent. Recently, Brill and Russell have studied the thermal decomposition of AN by rapid-scan FTIR/thermal profiling technique and detected HNO_3 (g) and NH_3 (g), above 180^{0}C, formed by proton transfer process [13]. Shortly thereafter, AN begins to be detected in the gas phase, probably as a result of the recombination of HNO_3 (g) and NH_3 (g).

$$NH_4NO_3\,(s) \rightarrow (HNO_3 + NH_3)_{adsorbed} \rightarrow HNO_3\,(g) + NH_3\,(g) \rightarrow NH_4NO_3\,(g)$$

Thermal decompposituion studies of bulk AN conducted over the past several years revealed the difficulty to distinguish between reactions in the gas and condensed phases. Recent studies on the thermal stability of AN mixtures with variety of organic explosives suggest that AN or its decomposition products desensitize them [14]. The sensitiveness of AN towards impact and friction is relatively low.

4.3. Hydrazinium Nitrate (HN)

HN exists in two crystalline forms of different melting temperatures: the stable form melt at 70.7 ^{0}C, while the labile form melts at 62.1 ^{0}C. HN is practically stable at 75 ^{0}C and slightly less hygroscopic than AN. The significant difference of HN compared to other nitrate salts is the formation of NH_3, which enhances the explosive sensitivity [13]. HN is sensitive to impact and friction resembling tetryl.

4.4. Guanidinium Nitrate (GN)

GN exists as crystals melting at 215-216 ^{0}C. It represents the most important intermediate product for the commercial manufacture of both nitro and amino guanidines. Moreover, GN can be used as an ingredient of propellants and also a component of fusible mixtures containing AN. High rate thermolysis of 1,2,3-triaminoguanidinium nitrate (TAGN) liberates predominantly HNO_3 (g) and NH_3 (g) as the initial products [15]. The second stage decomposition products are NO_2, N_2O and HCN, which are then followed by third stage products CO_2, H_2O, NO and probably NO_2.

4.5. Urea and Thiourea Nitrates

Urea nitrate (UN) is prepared by treating urea with dilute nitric acid (34%). Due to its instability, its practical applications are limited. UN decomposes into CO_2, N_2O, AN and urea at 140 ^{0}C and at 180 ^{0}C decomposition is rather violent. UN is impact and friction insensitive. Thiourea nitrate is found to be a weak explosive.

4.6. Methyl Ammonium Nitrates

Methyl ammonium nitrates are used as an ingredient of fusible explosives. Jain *et al* have prepared mono-, di- and trimethyl ammonium nitrates by neutralizing the aqueous solution of the corresponding amine with nitric acid [16]. Monomethylammonium nitrate (MAN) showed a structural transformation at 86 ^{0}C whereas, trimethylammonium nitrate showed two transformations, one at 86 ^{0}C and the other at 136 ^{0}C. The decomposition temperature of these compounds increases in the order: $CH_3NH_3NO_3 > (CH_3)_2NH_2NO_3 > (CH_3)_3NHNO_3$. Thermolysis of these salts usually involves proton transfer and the overall decomposition process can be represented as:

$$(CH_3)_nNH_{4-n}NO_3 \rightarrow (CH_3)_nNH_{3-n} + HNO_3 \text{ Products}$$

The chief decomposition products are N_2, NO, NO_2 and H_2O. However, MAN also yields CH_3NH_2 and NH_3 by methyl group transfer [16, 17].

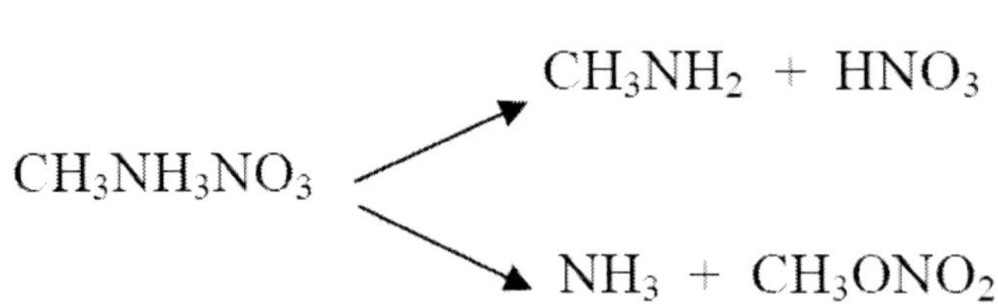

Tetramethyl ammonium nitrate (TAMN) has no explosive properties. However, it is being used as an ingredient of explosive mixtures blended with cyclonite or incorporated in fusible mixture with AN. TAMN undergoes decomposition as follows:

$$(CH_3)_4NNO_3 \rightarrow (CH_3)_3N + CH_3NO_3 \rightarrow \text{Products}$$

Methyl nitrate was found to decompose above 200^0C forming formaldehyde, methanol, water, carbon monoxide, nitric oxide and nitrogen dioxide [18]. The decomposition of TMAN [19] proceeds via dissociation into trimethyl amine and methyl nitrate involving a methyl group transfer. The activation energy for decomposition has been determined by isothermal TG and formation of trimethyl amine was monitored by spectroscopy.

4.7. POLYAMMONIUM NITRATES

Polyammonium nitrates have been used as a component of explosives for many years. Mc Kenny *et al* have synthesized a series of selected organodi- and poly ammonium nitrate salts and characterized by 1HNMR, DTA, DSC, FTIR and mass spectrometry [20].

These salts (table 2) were prepared by reacting the base with 70% nitric acid in cold 95% ethanol. The acid was added dropwise to the cold ethanolic solution of the respective amine with stirring, and the low temperature minimizes the formation of ethyl nitrate. Impact sensitivity of these nitrates was found to be related to oxygen balance (OB). The thermolysis of these salts has been reported to be quite complex.

Fast thermolysis patterns and pre-ignition reactions of EDD, BDD, HDD and methanediammonium dinitrate (MDD) have been described by Russell and Brill [21]. EDD decomposes to HNO_3 and NO_2 at about 275^0C and NH_3 appear at about 300^0C as a result of C-N bond fission and proton transfer [22]. The reaction of HNO_3 with NH_3 in the gas phase produces NH_4NO_3 aerosol at about 310^0C. The reduction of HNO_3/NO_2 to NO and the oxidation of hydrocarbon residue to CO/CO_2 occurs at about 330^0C.

PDD undergoes a solid-solid phase transition at 62^0C and melts at 126^0C without decomposition [23]. Evolution of HNO_3 (g) occurs at 280^0C followed by the oxidation-reduction products of HNO_3 and the alkyl amine. Mass spectrometric investigations on the initial stages of thermal decomposition of BDD reveal the formation of the corresponding amine and nitric acid at 160-180^0C. The nitric acid partially decomposes to produce nitrogen dioxide.

The fast thermolysis patterns [24] of pentaerythrityltetrammoniumtetranitrate (PETN) exhibit the initial occurrence of the cleavage of N-H bond to form HNO_3. Some C-N bond cleavage also takes place producing NH_3 and NH_4NO_3 (g). Partial oxidation of NH_3 by HNO_3 alongwith the decomposition of HNO_3 leads to the formation of NO_2, N_2O and H_2O. PIPZN yields initially HNO_3, but then generates a significant amount of N, N`-dinitrosopiperazine [25]. Cubylammonium nitrate (CUBAN) and cubane-1, 4-diammoniumdinitrate (CUBDAN) undergo thermal decomposition in the solid state without melting whereas; most of the other alkylammonium nitrate salts melt prior to decomposition [26]. CUBDAN is the only alkyldiammonium dinitrate salt found to sublime during fast thermolysis.

Table 2. Molecular formula, OB and melting point of organodi - and polyammonium nitrate salts

Compound name	Acronym	Molecular formula	OB	Melting point (^{0}C)
Homologues of EDD				
1. Ethylenediammonium dinitrate	EDD	$C_2H_{10}N_4O_6$	-1.07	188
2. 1,3-Propanediammonium dinitrate	PDD	$C_3H_{12}N_4O_6$	-3.00	126
3. 1,4-Butanediammonium dinitrate	BDD	$C_4H_{14}N_4O_6$	-4.67	139
4. 1,6-Hexanediammonium dinitrate	HDD	$C_6H_{18}N_4O_6$	-7.43	109
N-Sustituted EDD				
5. N-Methylethylenediammonium dinitrate	MEDD	$C_3H_{12}N_4O_6$	-3.00	78
6. N-Ethylethylenediammonium dinitrate	EEDD	$C_4H_{14}N_4O_6$	-4.67	86
7. N-Isopropylethylenediammonium dinitrate	IPEDD	$C_5H_{16}N_4O_6$	-6.13	94
N,N`-Substituted EDD				
8. N,N`-Dimethylethylenediammonium dinitrate	s-DMEDD	$C_4H_{14}N_4O_6$	-4.67	130
9. N,N`-Dimethylethylenediammonium dinitrate	u-DMEDD	$C_4H_{14}N_4O_6$	-4.67	52
10. N,N`-Diethylethlenediammonium dinitrate	s-DEEDD	$C_6H_{18}N_4O_6$	-7.43	193
N,N,N`-Substituted EDD				
11. N,N,N`-Trimethylethylenediammonium dinitrate	TRMEDD	$C_5H_{16}N_4O_6$	-6.13	93
12. N,N-Dimethyl-N`-ethylethylenediammonium dinitrate	DMEEDD	$C_6H_{18}N_4O_6$	-7.43	101
13. N,N,N`-Triethylethylenediammonium dinitrate	TREEDD	$C_8H_{22}N_4O_6$	-9.62	86
N,N,N`,N`-Substituted EDD				
14. N,N,N`,N`-Tetramethylethylenediammonium dinitrate	TMEDD	$C_6H_{18}N_4O_6$	-7.43	228
15. N,N,N`,N`-Tetraethylethylenediammonium dinitrate	TEEDD	$C_{10}H_{26}N_4O_6$	-11.40	140
Miscellaneous				
16. Diethylenediammonium dinitrate	DETN	$C_4H_{16}N_6O_9$	-2.05	150
17. Triethylenetetrammonium tetranitrate	TETN	$C_6H_{22}N_8O_{12}$	-2.51	229
18. Tetraethylenepentammonium pentanitrate	TEPP	$C_8H_{28}N_{10}O_{15}$	-2.78	196
19. Piperazinium dinitrate	PIPZD	$C_{412}N_4O_6$	-3.77	224

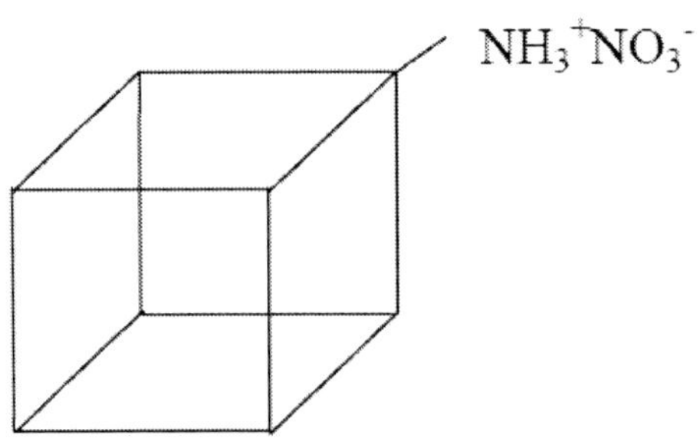

(CUBAN)

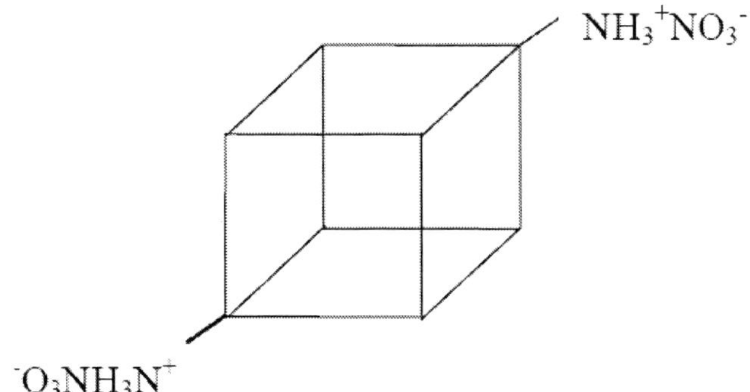

CUBDAN

The fast thermolysis of primary alkanediammonium dinitrate salts $[H_3N(CH_2)_nNH_3](NO_3)_2$ (where n = 1-4 and 6) is potentially intriguing because hydrogen bonding can cause intramolecular cyclization [27, 28] of the longer chain diamines (n = 4, 6). The behaviour of N-methylethylenediammonium dinitrate (MEDD), N, N, N`, N`-tetramethylethylenediammonium dinitrate (TMEDD), N, N, N`-trimethylethylenediammonium dinitrate (TRMEDD), N,N`-dimethylethylenediammonium dinitrate (s-DMEDD) and N, N`- dimethylethylenediammonium dinitrate (u-DMEDD) are more closely resemble the behaviour of alkylammonium mononitrate salts in terms of the temperature at which HNO_3 is released [29]. The first decomposition step of most primary and secondary alkylammonium nitrate salts is the production of HNO_3 (g).

4.8. Anilinium Nitrates

The thermal stability of ring (mono) substituted anilinium nitrates (RSAN) was investigated by Jain *et al* [30]. The decomposition temperature of meta- and para- RSAN was found to be linearly related to the Hammett substituent constant (σ). The activation energy for decomposition obtained by isothermal TG, increases with the increasing basicity of the corresponding arylamine. The primary step in the decomposition of RSAN salts is proton abstraction by the nitrate anion from the anilinium ion. Recently, Singh and Kapoor [31] have studied the kinetics of thermolysis of RSAN by ignition delay measurements. The ignition temperature and activation energy were found to be linearly related with the σ, pK_a and OB of the corresponding nitrate salts.

The mechanism of thermolysis of these salts has also been proposed based on the formation of reaction intermediates during decomposition. A reaction (Scheme 2), accounting for the products was also proposed that involves proton transfer leading to the formation of an arylamine and HNO_3 and oxidation of the arylamines by decomposition products of HNO_3. The oxidation-reduction reactions near the surface of the thermolysing RSAN were reported to be responsible for the decomposition.

The thermolysis of dimethylanilinium nitrates (DMAN) has also been carried out by TG, DTA, ignition delay, impact and friction sensitivity measurements. Thermolysis of DMAN involves complex reaction paths, and proton transfer seems to control the overall thermolysis. Electron donating groups ($-CH_3$) reduce the sensitivity of the energetic compounds within a class. A reaction Scheme 2 has been proposed where an arylamine and HNO_3 are formed due to proton transfer. The evolution of NH_3 (C-N bond heterolysis) has also been suggested (Scheme 3) prior to the oxidation of aryl nucleus by NO_2, HNO_3 and /or its decomposition products, and ultimately gaseous products, are formed.

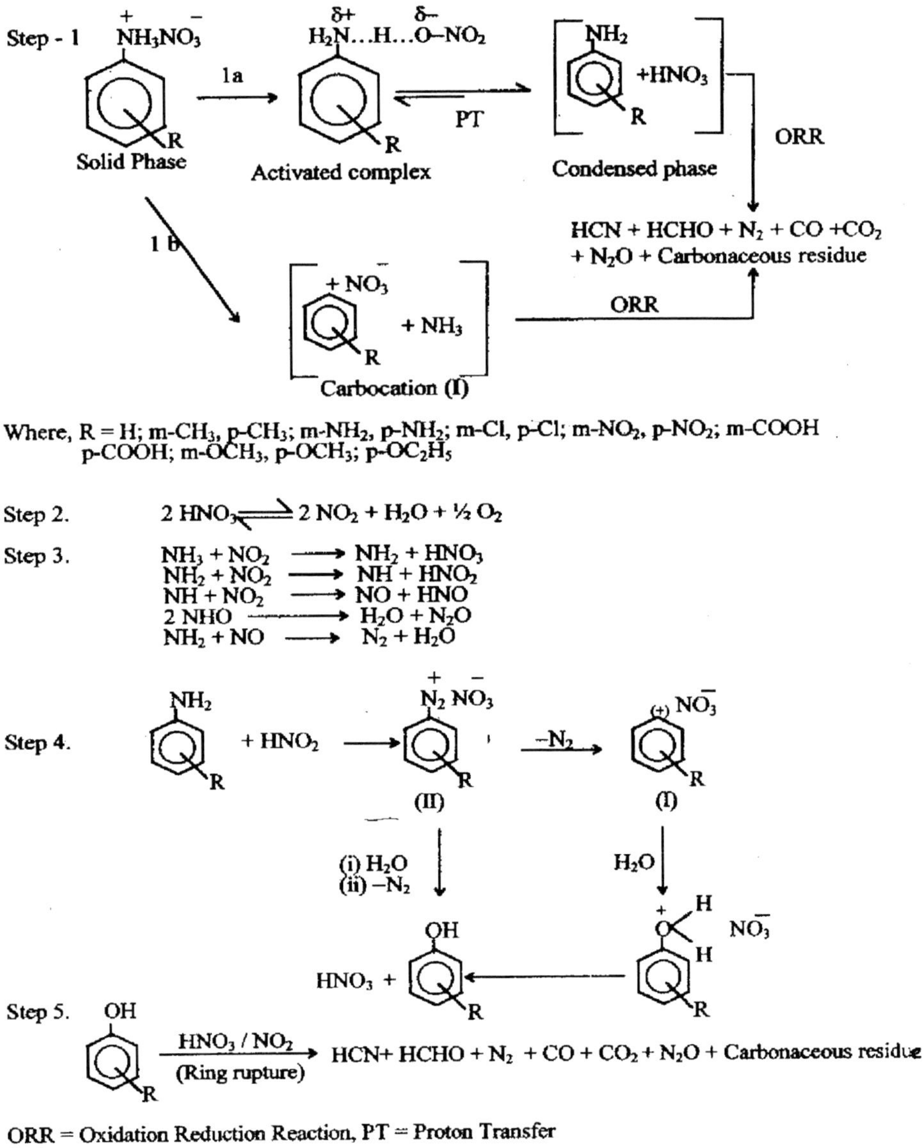

Scheme 2. Schematic representation of the thermal decomposition pathways of dimethylanilinium nitrates.

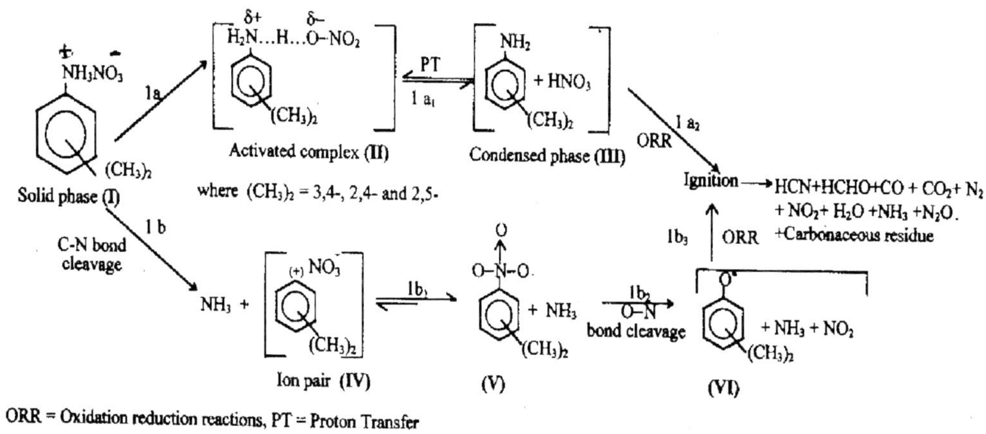

Scheme 3. Schematic representation of the thermal decomposition pathways of ring substituted arylammonium nitrates.

4.9. Conclusion

AN is hygroscopic and undergoes room temperature phase transition. It melts at 169.5^{0}C after the initial polymorphic transitions. The mechanism of thermolysis of AN is highly dependent on the experimental conditions. Its decomposition starts with proton transfer from NH_4^+ to NO_3^-. GN, methyl ammonium nirates, polyammonium nitrates and anilinium nitrates have also proton transfer thermolysis. Urea nitrate is unstable and decompose into CO_2, N_2O, AN and urea at 140^{0}C.

References

[1] G. P. Sutton, Rocket Propulsion Elements, 50th Ed. Wiley, New York, (1986) pp 296.

[2] C. Oommen and S. R. Jain, *J. Hazard. Mater.,* A67 (1999) 253.

[3] K. Krishnan, R. Rajeev, K. G. Kannan and K. N. Ninan, *Proc., 13th National Symp. On Ther. Anal.* THERMANS – 2002, BARC, Mumbai p 148.

[4] A. C. Mc Laren, *Rev. Pure Appl. Chem.,* 12 (1962) 54.

[5] S. B. Hendricks, E. Posnjak and F. C. Kracek, *J. Am. Chem. Soc.,* 54 (1932) 2766.

[6] N. R. Brown,and A. C. Mc Laren, *Proc. Roy. Soc.,* 266 (1962) 329.

[7] J. S. Ingman, G. J. Kearly and S. F. A. Kettle, *J. Chem. Soc. Farady Trans.,* 178 (1982) 1817.

[8] C. N. R. Rao, B. Prakash and M. Natarajan, Crystal Structure, Transformation in Inorganic nitrites, Nitrates and Carbonates, National Bureau of Standards, NSRDS-NNS 53 May 1975.

[9] D. R. Lide and H. V. Kehiain, CRC Handbook of Thermophysical and Thermochemical data: Eds. CRC Press: Boca Raton, 1994.

[10] Sunders, *J. Chem. Soc.,* (1922) 198.

[11] M. S. Shah and T. M. Oza, *J. Chem. Soc.,* (1932) 725.

[12] G. Feick nand R. M. Hainer, *J. Am. Chem. Soc.,* 76 (1954) 5860; Nature 173 (1954) 1188.
[13] T. P. Russell and T. B. Brill, *Combust. Flame,* 76 (1989) 393.
[14] J. C. Oxley, *J. Phys. Chem.,* 98 (1994) 3893.
[15] Y. Oyumi and T. B. Brill, *J. Phys. Chem.,* 89 (1985) 4325.
[16] S. R. Jain, M. V. Rao and V. R. Pai Verneker, *Propell. Explos.,* 3 (1978) 83.
[17] P. Breisacher, H. H. Takimoto, G. S. Denault and W. A. Hicks, *Combust. Flame,* 14 (1970) 393.
[18] A. Appian, O. Todes and Yu Khantor, J. *Phys. Chem.* (USSR) 8 (1936) 866.
[19] P. R. Nambiar and S. R. Jain, *Thermochim. Acta,* 9 (1974) 295.
[20] R. L. McKenny Jr., S. R. Struck, R. A. Hildreth and J. A. Fryling, *J. Energ. Mater.,* 5 (1987) 1.
[21] T. P. Russell and T. B. Brill, Propell. *Explos. Pyrotech.,* 15 (1990) 77.
[22] D. G. Patil and T. B. Brill, *Combust. Flame,* 87 (1991) 145.
[23] R. Chen, T. P. Russell, A. L. Rheingold and T. B. Brill, *J. Cryst. Spect. Res.,* 21(1991) 167.
[24] Y. Oyumi, T. B. Brill, A. L. Rheingold and C. Lowe-Ma, *J. Phys. Chem.,* 89 (1985) 2309.
[25] T. P. Russell, T. B. Brill, A. L. Rheingold and B. S. Haggerty, *Propell. Explos. Pyrotech., 15* (1990) 81.
[26] T. P. Russell and T. B. Brill, *Propell. Explos. Pyrotech.,* 16 (1991) 27.
[27] R. Yamdangni and P. Kebarle, *J. Am. Chem. Soc.,* 95 (1973) 3504.
[28] D. H. Aue, H. M. Webb and M. T. Bowers, *J. Am. Chem. Soc.* 95 (1973) 2699.
[29] T. P. Russell and T. B. Brill, *Propell. Explos. Pyrotech.,* 15 (1990) 123.
[30] S. R. Jain, M. V. Rao and V. R. Pai Verneker, *J. Chem. Soc., Perkin II* (1979) 406.
[31] G. Singh and I. P. S. Kapoor, *Combust. Flame,* 92 (1993) 283.
[32] G. Singh, I. P. S. Kapoor and S. M. Mannan, *Thermochim. Acta,* 262 (1995) 117.

Chapter 5

CHEMISTRY AND THERMOLYSIS OF POLYNITRO COMPOUNDS

ABSTRACT

This chapter includes the chemistry and thermolysis of different nitro compounds same as nitroguanidines, aliphatic and aromatic nitro compounds, homocyclic and heterocyclic nitro compounds and metal salts of nitro compounds.

5.1. INTRODUCTION

Nitro group is a powerful plosophoric group, capable of forming an energetic compound on introduction into a hydrocarbon. Mostly nitro compounds are prepared by nitration, which is carried out either directly introducing nitro group in place of hydrogen atom or by adding it to a double bond, or indirectly by introducing a group, which can readily be substituted/replaced by the nitro group. Most of the common explosives such as TNT, RDX, HMX, TATB, TNAZ, NTO etc. are the nitro compounds. A brief description of the various nitro compounds are given.

5.2. NITRATION

The following nitrating agents are frequently used for the direct introduction of nitro groups.

(1) Concentrated nitric acid
(2) Mixture of concentrated nitric and sulphuric acid (nitrating mixture).
(3) Alkali nitrates in the presence of sulphuric acid
(4) Dilute nitric acid
(5) Nitrogen dioxide
(6) A solution of nitrogen dioxide in sulphuric acid
(7) Nitrogen dioxide in presence of a catalyst.

The composition of a nitrating mixture should be chosen very carefully with regard to the compound being nitrated and the number of nitro groups to be introduced. Polynitro compounds can probably be formed only from paraffins of fairly high molecular weight [1].

5.3. NITROGUANIDINES

Nitroguanidines (NQ) is an important ingredient of flashless propellants and a potential insensitive explosive. NQ has two crystalline modifications called α and β- forms both belong to the orthorhombic crystal system [2]. Spectroscopic investigations [3, 4] on the structure of NQ and its derivatives exhibit their existence in nitramine form. The most important method of preparing NQ is the addition of guanidine nitrate to concentrated sulphuric acid.

During thermal decomposition of NQ besides the simple decomposition products like N_2O, NH_3, CO_2, HCN, urea, cyanogens, cyanamide, cyanic acid, polymeric products like melamine, ammeline, ammelide, cyanuric acid, melem, melam, melon and para cyanogens were also identified [5, 6]. Since, NQ is made up of two amino groups, one nitramine group and one carbon atom, which may be easily availed as an ideal element to form a desired explosive molecule [7]. The reaction between NQ with formaldehyde or urotropin form a number of polynitro heterocyclic explosive and its derivatives, which are valuable intermediates for preparing some high explosives [8].

The more energetic relative of NQ is N-nitro-N-(2,2,2-trinitroethyl)guanidine (TNENQ) and is a potential burn rate accelerator [9], two or three times greater than HMX. TNENQ mainly exists in nitramine form and is prepared by reacting NQ with formaline to produce methylolnitroguanidine, followed by the reaction of trinitromethane.

$$\underset{\textbf{NQ}}{NO_2NHC(NH)NH_2} \xrightarrow{CH_2O} NO_2NHC(NH)NHCH_2OH \xrightarrow{CH(NO_2)_3} \underset{\textbf{TNENQ}}{NO_2NHC(NH)NHCH_2C(NO_2)_3}$$

Thermolysis of TNENQ liberates NO_2, NO, N_2O, CO_2, CO, H_2O, HCN and HNCO and cause slight sublimation of the parent molecule. Impact sensitivity of NQ is very less than that of TNENQ since, the hydrogen bonding present in the NQ crystal lattice would absorb impact energy and thereby desensitize it [3].

5.4. ALIPHATIC NITRO COMPOUNDS

Aliphatic nitro compounds with nearly zero or even positive OB and high melting points are desirable explosives and also are useful oxidizers in solid propellants. Early work on synthesis and chemistry of nitroparaffins was reviewed by various workers [10-12], later an extensive survey on synthesis of aliphatic polynitro compounds and their derivatives was made by Noble *et al* [13]. New process to make polynitro aliphatic derivatives have been discovered, in which an additional electronegative moiety has been incorporated into the molecule at the methyl group. This class of compounds have the general formula: $RC(NO_2)_2X$, where X = NO_2 (trinitromethyl [14]), F (fluorodinitromethyl [15], CN (cyanodinitromethyl [16]) and N_3 (azidonitromethyl [17]). The incorporation of an azide group have obvious energy benefits, because of its high energy content (80 kcal/mol). Lower polynitro aliphatic compounds may be obtained [18] by direct vapour phase nitration of aliphatic hydrocarbons either with HNO_3 vapour or with NO_2 and they form powerful explosive mixtures with combustible substances. Higher polynitro aliphatic compounds are normally prepared [19] by reacting ammonia with trinitroethanol. Introduction of multinitro groups can be achieved by Double Michael Reaction [13].

5.5. Aromatic Nitro Compounds

Aromatic nitro compounds as explosives are of greatest practical importance [20, 21] are obtained mostly either by the nitration of corresponding aromatic compound or by the oxidation of corresponding amines. Presence of even one nitro group is sufficient to increase the ease of thermal decomposition of aromatic compounds. Nevertheless, aromatic compounds which have two or more nitro groups, exhibit distinctly marked explosive properties. Zeman *et a.l* [22, 23] have correlated the thermal decomposition kinetics of polynitro aromatic explosives at lower temperatures with molecular structure, heat of explosion, detonation characteristics and thermal stability.

Several polynitro aromatic compounds undergo thermal decomposition by simple bond scission or rearrangements of the substituents on the ring [24]. Mostly, C-NO_2 homolysis to form NO_2 and an aromatic radical, is a dominant decomposition channel and it appears to be a major early thermal decomposition reaction at higher temperatures. NO_2 gas was rarely detected as a decomposition product and the evolution of large quantity of NO has been reported [24]. The next stage of thermal decomposition is propagation reactions, which involve exothermic oxidation and reduction that lead to lower molecular weight and thermodynamically stable products.

5.5.1. Nitrobenzenes

m-Dinitrobenzene (DNB) can be prepared [25] either by the nitration of benzene or by the oxidation of nitroaniline. The explosive decomposition products DNB are CO_2, CO, H_2, N_2, CH_4, HCN, C_2H_2, NH_4HCO_3 and C and its VOD is 6100 m/s. Among the various isomers of trinitrobenzene, 1,3,5-trinitrobenzene (TNB) is the chemically most stable one. It has high stability, great explosive power and less sensitivity to impact and friction. TNB can be obtained [26] by nitrating DNB by a mixture of HNO_3 and oleum. 1,2,4,6-Tetranitrobenzene is an impact sensitive exlosive, readily decomposes to picric acid in presence of moisture. Hexanitrobenzene can be prepared [27] by oxidizing the amino group of pentanitroaniline with H_2O_2 in H_2SO_4. Tetra- and penatnitrobenzene can also be obtained [28] by the oxidation of their corresponding nitroanilines using peroxydisulphuric acid.

5.5.2. Trinitrobenzene Derivatives

Trinitrobenzene derivatives have long been important in technological applications. Their analytical and $h_{50\%}$ data are compiled in table 1. 1-amino-2,4,6-trinitrobenzene (MATB) (compound no. 1) is prepared by the successive nitration of aniline or by heating picric acid with urea. 1,3-diamino-2,4,6-trinitrobenzene (DATB) (2) is mainly prepared from stiphnic acid and is used in high explosive compositions.1,3,5-Triamino-2,4,6-trinitrobenzene (TATB) (3) can be readily obtained from aniline and it has qualified as heat resistant explosive. The unique structure of TATB confers properties which include high density and extremely insensitive to impact and shock. Moreover, the introduction of amino groups to these molecules, adds sufficient energy to the crystal lattice and increase the melting point. The initial decomposition mechanism of MATB, DATB and TATB below 500^0C is more complicated, since an NH_2 group ortho to $-NO_2$ in an aromatic ring will cyclize on heating to form furazan and furoxan derivatives. Their explosive sensitivity was found to be in the order: MATB>DATB>TATB. However, the desensitizing influence of NH_2 group is due to its electron donating and /or hydrogen bonding characteristics.

2,4,6-(trinitrophenyl)methylnitramine (Tetryl) (4), 3-aminotetryl (3AT) (5) and 3,5-diaminotetryl (3,5DAT) (6) are practical explosives. Tetryl (4) can be prepared by the nitration of N-methylaniline. Hutchinson *et al.* [29] have prepared (5) and (6) from their corresponding chlorotetryl by the displacement of halogen by ammonia. Their explosive sesitiveness is in the order: (6)>(5)>(4). Koppes *et al.* [30] have prepared 1,3,5-trifluorotrinitrobenzene (TFTNB) (7) by the direct nitration of 1,3,5-trifluorobenzene. Difluoropicramide (DFP) (8) and fluorodiaminotrinitrobenzene (FDATB) (9) have been prepared from compound (7). He compound (7), (8) and (9) are very good precursors to synthesis a large number of TNB derivatives. TFTNB reacted with formamide, in presence of THF to give the formamide derivatives (10) and (11) and with urea to give ureido derivative (12). TFTNB on reflexing with formamide in p-dioxane gave (13). The compounds (14), (15) and (16) can be obtained from FDATB by the direct displacement of fluorine. The aromatic nitriles (17) and (18) were hydrolysed by conc. H_2SO_4 to the corresponding amides (19) and (20) respectively, and 70% aqueous H2SO4 converted them to the carboxylic acid (21).

The compounds (8), (10) and (11) were hydrolysed to the substituted styphnic acids (22), (23) and (24) respectively. The starting materials (7), (8) and (9) reacted with fluorodinitroethanol to give the fluorodinitroethoxy derivatives (25), (26) and (27) and reacted with fluorodinitroethylamine to give the fluorodinitroethylamino derivatives (28), (29) and (30) respectively. Reddy *et al.* [31] have undertaken studies on the kinetics of decomposition of picryl azide (31) by DTA, DSC and isothermal TG techniques. Picryl azide melts (83-93^0C) and on decomposition looses N_2(g) to yield 4,6-dinitrobenzofurazan oxide, which decomposes to give gaseous products along with carbonaceous residue. The N-O bond cleavage is reported as the rate controlling step. The replacement of a nitro group in a polynitro compound by fluoroxano groups can result in the increase of density and VOD. Accordingly, Naixing *et al.* [32] have synthesized N,N`-bis(2,4-dinitrobenzofuroxan)1,3,5-trinitro-2,6-diaminobenzene as an explosive with high density (1.92 g/cm^3) and high VOD (8574 m/s) by taking 1,3-diaminobenzene and 1,3-dichloro-TNB as the starting materials.

Table 1. Structure, melting point, density and impact sensitivity data of trinitrobenzene derivatives

Structure	Compound No.	Substituents	m.p. (^{0}C)	Density (g/cm^3)	$h_{50\%}$ (cm)
1	2	3	4	5	6
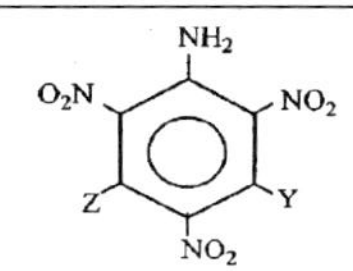	1 2 3	Y, Z = H Y = NH_2; Z = H Y, Z = NH_2	188 286 350	1.760 1.837 1.938	177 260 490
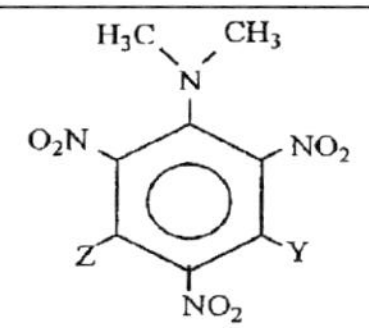	4 5 6	Y, Z = H Y = NH_2; Z = H Y, Z = NH_2	129 188(d) 186(d)	1.710 -- --	-- -- --
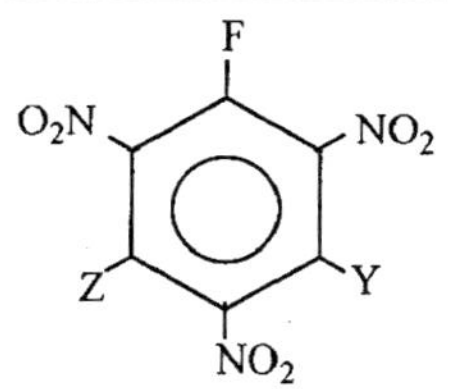	Y, Z = F Y = F; Z = NH_2 Y, Z = NH_2 Y = F; Z = NHCHO Y, Z = NHCHO Y = F; Z = $NHCONH_2$	82 118 223 179 238 164	2.00 1.939 1.946 -- -- --	-- 62 176 -- -- --	7 8 9 10 11 12
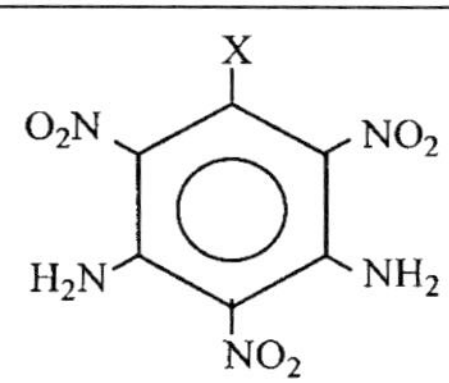	X = NHCHO X = $NHCONH_2$ X = $NHC(NH_2)$=N-NO_2 X = $NH(CH_2)_2NH_2$	229 >350 184 305	1.83 1.90 1.65 1.80	226 320 466 189	13 14 15 16

Table 1. (Continued)

Structure	Compound No.	Substituents	m.p. (^0C)	Density (g/cm^3)	$h_{50\%}$ (cm)
1	2	3	4	5	6
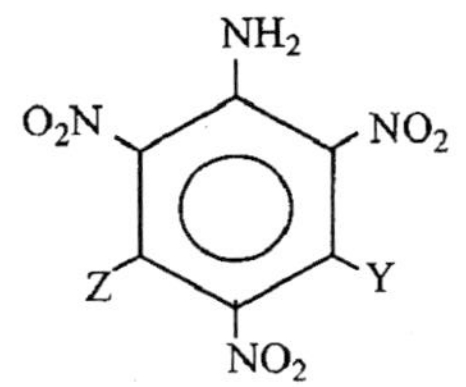	17	Y = CN; Z = NH_2	221	1.83	184
	18	Y, Z = CN	179	1.76	93
	19	Y = $CONH_2$; Z = NH_2	290	1.88	221
	20	Y, Z = $CONH_2$	>340	1.85	168
	21	Y = COOH; Z = NH_2	240	1.86	>320
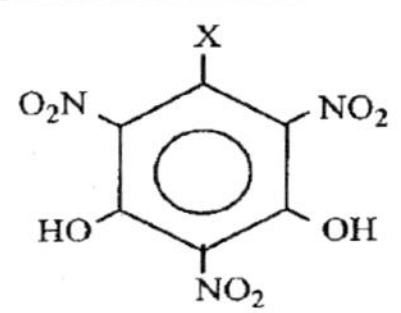	22	X = NH_2	234	1.88	324
	23	X = NHCHO	174	1.71	74
	24	X = $NHCONH_2$	164	1.81	36
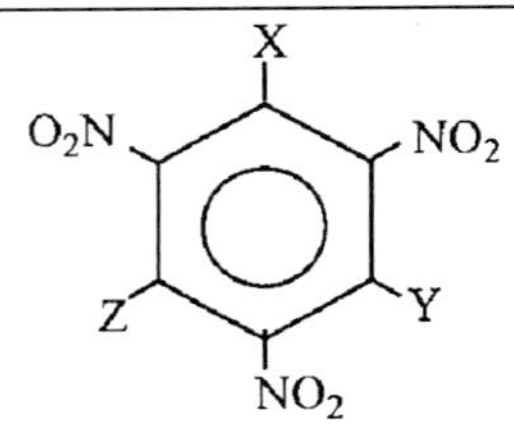	X = $OCH_2CF(NO_2)_2$; Y, Z = NH_2	190	1.82	568	25
	X, Y = $OCH_2CF(NO_2)_2$; Z = NH_2	140	1.79	300	26
	X, Y, Z = $OCH_2CF(NO_2)_2$	170	1.77	149	27
	28	X = $NHCH_2CF(NO_2)_2$; Y, Z = NH_2	198	1.82	493
	29	X, Y = $NHCH_2CF(NO_2)_2$; Z = NH_2	145	1.81	212
	30	X, Y, Z = $NHCH_2CF(NO_2)_2$	182	1.75	220
	31	X = N_3; Y, Z = H	83	--	--

5.5.3. Nitrotoluenes

The specific feature of nitration of toluene, is the presence of methyl group greatly facilitates the nitration as compared with benzene. Out of the various nitro derivatives of toluene, 2,4,6-trinitrotoluene (TNT) achieves its special position, because of its low hygroscopicity, high compatibility, melting point, relatively low sensitivity to impact and friction, good thermal stability and high power during explosion. TNT is normally prepared by the nitration of toluene by successive introduction of nitro groups. Its decomposition products are CO, CO_2, H_2, H_2O, N_2, NH_3, HCN, CH_4, C_2H_2, C_2N_2 and C. The evaporation of TNT starts above 200^0C and exothermic decomposition occurs at 300^0C with the involvement of some partly oxidized polymeric radicals as heat sensitizers. Below 770^0C, the initiation chemistry is dominated by oxidation reactions of the methyl group and above 770^0C, C-NO_2, homolysis dominates the initiation process. Tetranitrotoluene is more resistant to hydrolytic action of water and can not be obtained by the direct nitration of trinitrotoluene. Recently, Wilson *et al.* [33, 34] have developed new synthetic routes, which involve peroxydisulphuric acid oxidation of trinitrotoluidines and hypophosphorous acid reduction of diazonium salts derived from tetranitrotoluidine.

5.5.4. Nitrophenols

Among the three isomers of mononitrophenol, ortho and para are readily formed, when phenol is nitrated with dil. HNO_3. The meta isomer is usually prepared by diazotization of m-nitroaniline. Out of the six isomers of dinitrophenol (2,3-; 2,4-; 2,5-; 2,6-; 3,4- and 3,5-), 2,4- and 2,6-isomers may be obtained by the nitration of phenol and all other isomers are prepared by indirect methods. 2,4,6-Trinitrophenol (Picric acid) is of the greatest importance and is prepared by the nitration of 2,4-dinitrophenol. Its thermal decomposition can be represented as follows:

$$16C_6H_3O(NO)_3 \rightarrow 32CO_2 + 40CO + 4CH_4 + 2H_2 + 3C_2H_2 + 3HCN + 4.5H_2 + 21.5N_2 + 2NH_4HCO_3 + 9C$$

The impact and friction sensitivities of picric acid are higher than that of TNT and its VOD was reported to be in the range 4020-7504 m/s. 2,3,4,6-Tetranitrophenol was obtained by reacting hydroxylamine with dinitroresorcinol. It is more sensitive to impact than tetryl.

5.5.5. Nitrostilbenes and Nitrobibenzyls

4,4`-Dinitrostilbene was prepared by warming p-nitrotoluene in alkaline alcoholic solution. All the three possible isomers of 2,4,4`-trinitrostilbene can be obtained by heating o-, m- or p-nitrobenzaldehyde with 2,4-dinitrotoluene. 2,2`,4,4`6,6`-Hexanitrostilbene (HNS) is a heat resistant explosive, used as a crystal modifying additive and also in explosive boosters. The thermolysis of HNS and its derivatives have been studied by isothermal DSC. Zen-guo *et al.* [35, 36] described the synthesis of 3,3-bis(HNS) based on the Ullmann reaction by reacting 3-chloro-HNS with hydrazine. Synthesis of 3,3-disubstituted polynitrostilbenes having substituents with increasing electron donating ability (MeN>MeO>Cl) have also been described. Tetra nitrobibenzyl and 2,2,4,4`,6,6`-hexanitrobibenzyl can be prepared by treating nitrobenzyl halogenide with TNT.

5.6. Homocyclic Nitro Compounds

There is a considerable interest in the synthesis and thermolysis of polynitro homocyclic ("cage") compounds. These are relatively highly strained molecules that contain several $-NO_2$ substituents, and emerging as high-density energetic materials. Recently, Eaton *et al.* [37] have developed a systematic methodology for the syntheses of 1,3,5-trinitrocubane and 1,3,5,7-tetranitrocubane as superior high energy shock insensitive explosives. The thermal behaviour and detonation properties of D3-hexanitrotrishomocubane have been investigated and the explosive output calculations show that it is substantially more powerful explosive than TNT. Dave [38] has prepared 4,4`-dinitroadamentane-2,6-dione as an intermediate to make high energetic compounds. 1,4,6,9-Tetranitrodiamentane and 1,3,5,7-tetranitrodiamentane can be obtained by the oxidation of their corresponding tetraminodiamentane and adamentane by $KMnO_4$. These compounds have an OB of -121.7 and are useful as solid rocket fuel. 1,3,5,7-Tetranitroxyadamentane (TNOAD) has been prepared by the nitration of 1,3,5,7-tetrahydroxyadamantane. TNOAD has an excellent thermal and impact stability. A recent review by Merchand [39] describes briefly the synthesis and explosive performance of polynitropolycyclic cage explosives.

5.7. Heterocyclic Nitro Compounds

Heterocyclic nitro compounds may be aliphatic or aromatic in nature, depending upon the electronic constitution. These compounds represent explosives of higher performance compared with analogous aromatic systems, regarding their elemental composition, OB, density, standard heat of formation and VOD. Reactions of a polyatomic substituents with an ortho $-NO_2$ group in an aromatic ring are now widely used to synthesise many heterocyclic molecules [40-42]. In general, the reactions between $-NO_2$ and ortho related substituents profoundly affect the thermal decomposition chemistry and explosive properties of polynitroaromatic explosives. The proton transfer from $O\text{-}CH_3$ to $-NO_2$ is energetically favourable. Normally, decomposition of ortho substituted aromatic nitro compounds at lower temperatures favours cyclisation reactions whereas, higher temperatures favour homolysis of the $C\text{-}NO_2$ bond. Mc Kay and Wright [43] developed an overall synthetic route to prepare 2-nitramino-1,3-diazocycloalkenes by reacting alkyldiamines with nitroguanidine. This synthesis was found to be successful with diamines containing branched as well as straight chains.

5.7.1. Four-Membered Ring Compounds

Nitro substituted four-membered ring compounds possess high strain energy and are expected to have excellent explosive performance. It is more difficult to prepare these compounds by the direct cyclization of straight-chain intermediates. Azetidine based explosives such as 3,3-dinitroazetidine (DNAZ) and 1,3,3-trinitroazetidine (TNAZ) have occupied a significant place [44].

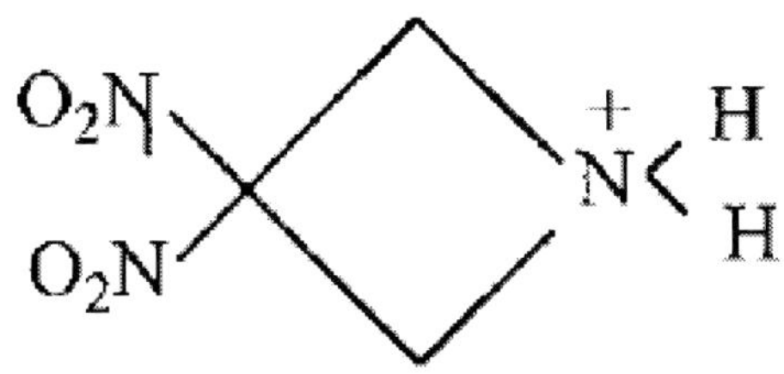

(DNAZ)

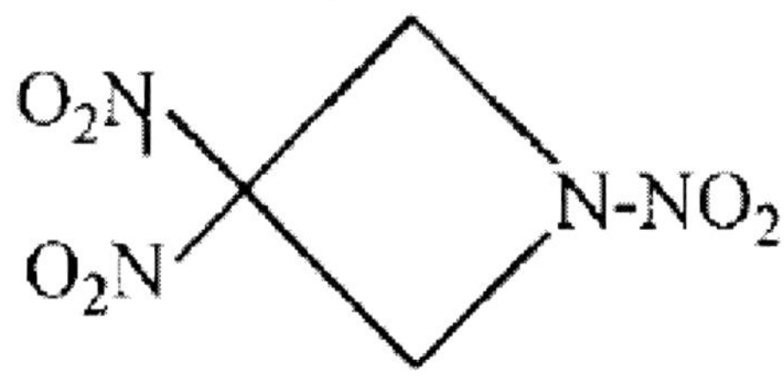

(TNAZ)

DNAZ has been prepared [45] from 1-t-butyl-3,3-dinitroazetine. The basicity of DNAZ allows the preparation of a large variety of salts, by introducing energetic anions like nitrate (DNAZ- Nitrate), dinitramidate (DNAZ-DN) and five membered heterocyclic aromatic ring namely, 2,4-dinitroimidazolate (DNAZ-DNI); 4,4`,5,5`-tetranitro-2,2`-biimidazolate (2DNAZ-N4BIM); 5-nitro-1,2,4-triazole-3-onate (DNAZ-NTO); 3,5-dinitro-1,2,4-triazolate (DNAZ-DNT). These DNAZ salts were prepared by either mixing the free base with the appropriate acid or by metathesis of 3,3-dinitroazetidinium trifluoromethanesulfonate with the ammonium salt of the acid. The density, thermal and $h_{50\%}$ data of these DNAZ salts are summarized in table 2.

Table 2. Density, thermal and impact data of 3,3-dinitroazetidine (DNAZ) salts

Compound name	Density (g/cm^3)	ΔH_f (kcal/mol)	DTA exotherm (^0C)	Impact sensitivity $h_{50\%}$ (cm)
DNAZ-Nitrate	1.764	-64±2	150	32
DNAZ-DN	1.791	-8±4	140	14
DNAZ-DNI	1.652	-18±12	160	54
2DNAZ-N4BIM	1.782	+5±15	150	71
DNAZ-NTO	1.757	-48±1	160	42
DNAZ-DNT	1.694	+20±2	150	37

5.7.2. Five Membered Ring Compounds

The C-nitro derivatives of five membered N- and N, O- heterocycles are of interesting energetic compounds. 2,5-Dipicryl-1,3,4-oxadiazole (DPO) was prepared as a thermally stable explosive by the cyclization of N,N-bis(2,4,6-trinitrobenzoyl)hydrazine. Polynitroimidazoles are generally obtained by the nitration of polyiodoimidazoles. 1,3-Dinitro-imidazolidine can be obtained by treating ethylene-dinitramine with formaldehyde. The synthesis of 3-picrylamino-1,2,4-triazole involves the condensation reaction of tetryl with aminotriazole. It yields stable explosives on nitration. 1,1-dinitro-3,3`-azo-1,2,4-triazole may be obtained from 3-amino-1,2,4-triazole by oxidation and reduction followed by nitration. It has more impact sensitivity and finds application as high energy propellant.

Among nitroazole explosives, 3-nitro-1,2,4-triazole-5-one (NTO) is emerging as a new candidate having high performance than RDX and HMX [46-50]. NTO is widely investigated for its thermal, impact, friction and shock sensitivities. Due to its acidic nature and ease of preparation from cheap starting materials, a large number of salts and derivatives of NTO is also known. As we have undertook considerable amount of research on the preparation and thermolysis of NTO and its salts, we will discuss the research on NTO in a separate chapter (Chapter 7).

5-Amino-3-nitro-1,2,4-triazole (ANTA) has received a fair attention as a new insensitive high explosive. Pevzner *et al.* synthesized ANTA for the first time by starting from 3-amino-1,2,4-triazole [51]. The recent method involves a four step process initiating from 3,5-diamino-1,2,4-triazole. Fast thermolysis studies on ANTA and NTO exhibit the formation of polymeric melon like cycline azine residues [52]. Thermal investigations by Brill *et al.* on six related nitramino-1,2,4-triazole and nitraminotetrazoles, reveal that the nitramine position of the molecule initiates the degradation [53].

Experimental studies shows that decomposition of ANTA occur in two steps [54] (figure 1). The first step is fast whereas second step is slow. First step corresponds to the removal of NO_2 followed by ring rupture and various condensed phase processes leading to the formation of CO_2, N_2O, HCN, HCNO etc. Kinetic analysis by applying model fitting method yield a single value of activation energy (~56 kJ/mol) for 36% mass loss. However, isoconversional method shows that activation energy changes with extent of conversion α (figure 4). The activation energy of explosion is found to be 48 kJ/mol. The available thermal sensitivity data reveal that it could be used as a good candidate where insensitivity and high performance are required.

The second step of thermolysis is due to the slow thermolysis of the left residue which is also supported by the presence of two DTG peaks (figure 1). Both, DTA and DSC curves (figures 1, 2 and 3) clearly indicates the exothermic nature of the first step decomposition as there is a exothermic peak. Second step of mass loss is slow and seems to be exothermic in nature. Although, there is no clear exothermic peak but curve line above base line indicates exothermic nature. Just before the first exothermic peak, there is an endothermic peak in DTA (both in N_2 as well as in air atmospheres, (figures 1 and 2) and DSC which is due to the melting of ANTA (m.p. 238 ^{0}C). After the melting, the compound decomposes exothermically giving a sudden weight loss. It is evident from simultaneous TG-DTG-DTA plot (figure 1) that the compound undergoes approximately 95.5% mass loss and very small amount (≈4.5%) of residue was left. T. B. Brill *et al* [53] have shown that the sublimation of ANTA occurs at all pressure. At lower pressure range, the degradation also occurs in condensed phase. Thus, the second gradual weight loss upto 650^0C may be due to sublimation and / or degradation, ring rupture, various polymerization reaction, bond breaking and making processes simultaneously.

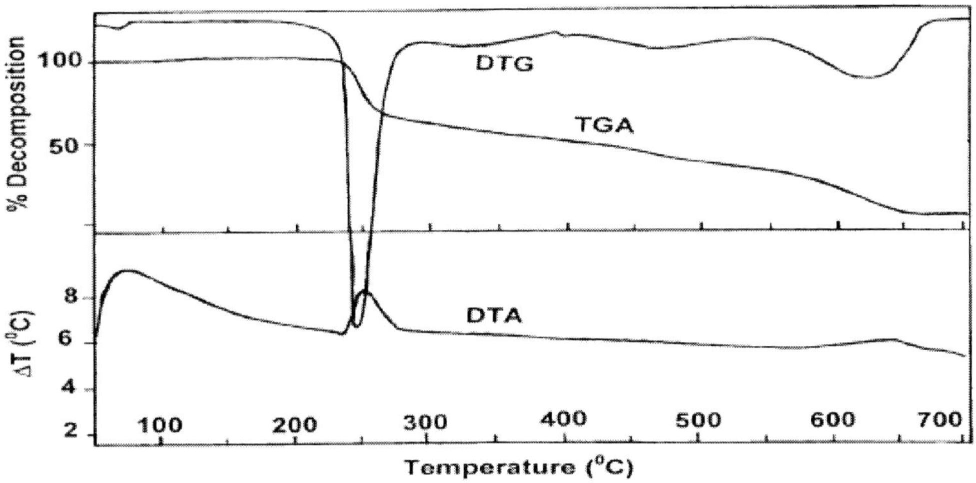

Figure 1. Simultaneous TG-DTG-DTA curve in N_2 atmosphere.

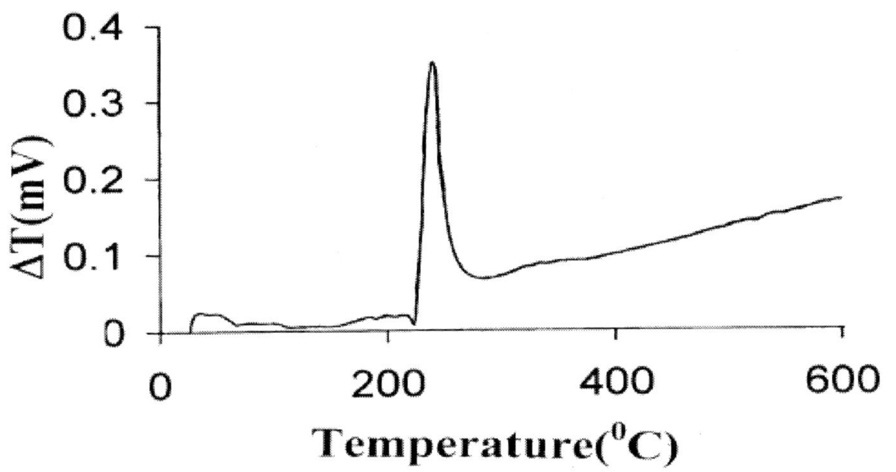

Figure 2. DTA curve of ANTA under flowing air atmosphere.

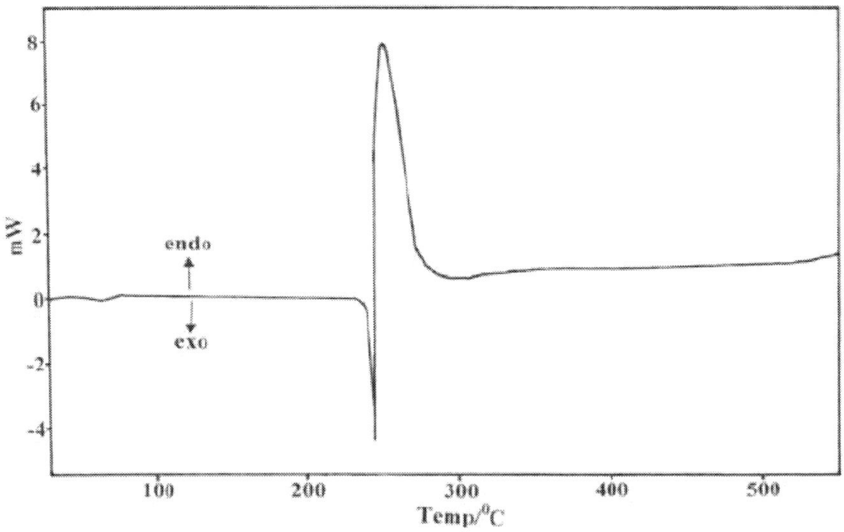

Figure 3. DSC curve of ANTA under flowing N_2 atmosphere.

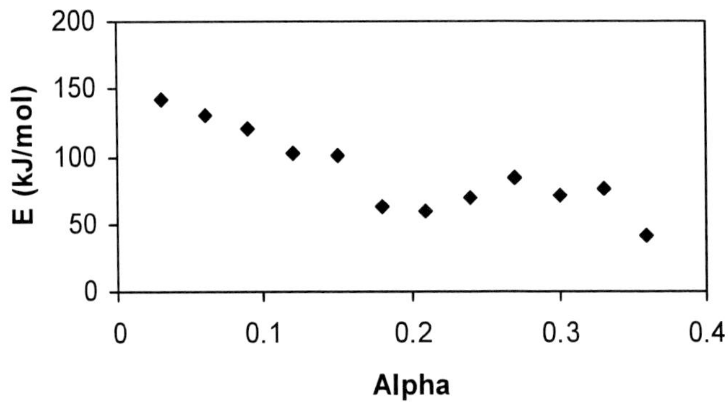

Figure 4. The dependencies of activation energy (E) on extent of conversion.

5.7.3. Six-Membered Ring Compounds

The most important ring system of this class is pyridine, which also show explosive nature, depending upon its substituents. Trinitropyridine and its derivatives namely: trinitropyridine-1-oxide (TNPyOx): 2,6-diazo-4-nitropyridine-1-oxide (DAzNPyOx); 2-methylnitramino-3,5-dinitropyridine (2-MNP); 4-methylnitramino-3,5-dinitropyridine (4-MNP); 2,6-bis(methylnitramino)3,5-dinitropyridine (2,6-BMNP); dinitrodipicrylpyridine (PYX). Their density, OB and thermal data are compiled in table 3.

Table 3. Density, OB and thermal data of six-membered, eight-membered and heterocyclic polynitro compounds

Compound	Density (g/cm)	OB	VOD (m/s)	DTA exotherm (^{0}C)	ΔH_f (kcal/mol)
TNPy	1.77	-37.4	8161	--	--
TNPyOx	1.86	-27.8	8369	--	--
DAzNOyOx	--	57.6	8532	--	--
2-MNP	1.71	-55.9	7585	190	--
4-MNP	1.71	-55.9	7500	210	--
2,6-BMNP	1.73	-47.9	7700	190	--
TNHHP	1.80	--	8360	--	17.5
TNHP	1.78	-39.8	8020	155	--
NMHP	1.79	-27.0	8210	190	--
RDX	1.80	-21.6	8763	--	60.2
HND	1.88	-16.3	8800	--	-17.0
HMX	1.91	-21.6	9124	--	60.5

The synthesis and initial thermal characterization of nitroderivatives of 3,5-dinitro-2,6-diaminopyridine have been carried out by Naixing *et al.* [55]. 2,6-Bis(picrylamino)-3,5-dinitromyrimidine (2,6-BPNP) has been synthesized from 2,6-diaminopyridine and 2,4-dinitrofluorobenzene in isopropanol solution by heating at 50^{0}C for 30 min and at 80^{0}C for 3.5 hours. To increase the density of 2,6-BPNP, replacement of nitro groups by benzofuroxano groups has also been suggested, since the introduction of aminoheterocycles in explosive molecule can result in increase of density and VOD and decrease of impact sensitivity.

1,3,5,5-Tetranitrohexahydropyrimidine (TNHHP) has been prepared as an oxygen rich energetic compound by the direct nitrolysis of 1,3-diisopropyl-5,5-dinitrohexahydropyrimidine (I):

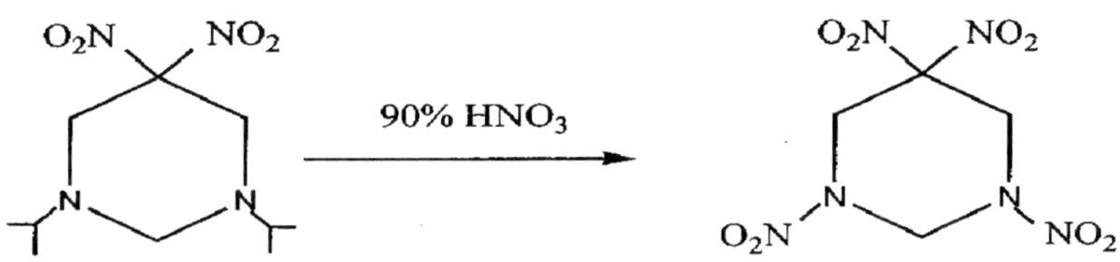

(I) (TNHPP)

The thermolysis of TNHHP has been recently investigated by Brill *et al.* [56]. Riter and Licht [57] have synthesized 1,3,5-trinitrohexahydropyrimidine (TNHP) by triple aminoalkylation of nitromethane with formaldehyde and tert-butylamine. TNHP on hydroxyl methylation and subsequent nitration converted it to 5-nitroxymethyl-1,3,5-trinitrohexahydropyrimidine (NMHP). TNHP, TNHP and NMHP possess pronounced explosive properties (table 4)

The most thoroughly studied system of triazine group is hexahydro-1,3,5-trinitro-1,3,5-triazine, which is commonly called RDX or cyclonite, which is a valuable high explosive (table 3). RDX is obtained from hexamethylenetetramine by nitration with N_2O_5 in 100% HNO_3 at -20 ^{0}C. It is binary energetic molecule, in which more than one energy-rich sites exist [58].

Early studies suggested that decomposition of RDX begins with N-N bond cleavage. However, a mass spectrometric study has indicated the possibility of C-N bond scission mechanism. The recent studies provide convincing evidence for N-N bond scission by the detection of N_2O_4 (dimer of NO_2). Some aspects of RDX decomposition is discussed in Chapter 8 wherein the thermolysis of plastic bonded explosives of RDX are discussed.

5.7.4. Eight Membered Ring Compounds

In recent years, a number of cyclooctane compounds containing two hetero (N) atoms (diazocines) have been synthesized as energetic oxidizers. Cichra and Adolph [59] have synthesized a series of polynitro- and nitronitrosoperhydro-1,5-diazocines. Frankel *et al.* have synthesized 1,3,3,5,7,7-hexanitrooctahydro-1,5-diazocine (HND) and characterized by thermal techniques (table 4). the replacement of $N\text{-}NO_2$ by N-NO decreases their impact sensitivity.

1,3,5,7-Tetranitro-1,3,5,7-tetraazacyclooctane commonly known as HMX (table 4) has recently occupied a key position in the field of military high-energy propellnts and explosives. At preset, it is used as the major ingredient of smokeless double-base propellants. Zeman *et al.* [60-62] have suggested a number of procedures to synthesise HMX from 1,5-diacetyl-3,7-dinitro-1,3,5,7-tetraazocyclooctane by nitrolysis. The thermal behaviour of HMX have been investigated extensively. The early works indicated that the thermal degradation of HMX proceeds by the predominant breakage of C-N bonds and NO_2 is eliminated. Recent investigations have shown that major part of the decomposition reactions of HMX occur in gas phase , which involves N-N bond ruptures. However, the formation of NO_2 can be hardly detected, since NO_2 react quite rapidly with formaldehyde, which has been recognized as the major product of HMx decomposition. Brill *et al.* revealed that CH_2O and N_2O are the main decomposition products at lower temperatures and HCN and NO_2 are formed at higher temperatures[63-65]. Frankel and Woolery have obtained 1-azidomethyl-3,5,7-trinitro-1,3,5,7-tetraazacyclooctane (AZTC) by replacing one NO2 group of HMX by an azidomethyl (CH_2N_3) group[66]. Thermal decomposition of AZTC involves C-N bond cleavage with the formation of N_2O and CH_2O.

5.8. Metal Salts of Nitro Compounds

Metal salts of nitro compounds have long been of interest as energetic additives and /or as burning rate modifiers in composite solid rocket propellants and in explosive compositions. Recently Singh *et al.* [67] have prepared Cu(II), Ag(I), Pb(II) salts of 2,4,N-trinitroanilino acetic acid (2,4,N-TNAAA) and 2,4,6-trinitroanilino acetic acid (2,4,6-TNAAA). The order of thermal stability of salts of 2,4,N-TNAAA is reported to be Ag>Pb>Cu, while for salts of 2,4,6-TNAAA, it is Cu>Ag>Pb.

Metal salts of NTO is currently receiving fair attention. Yi *et al.* [68] have prepared K, Cu and Pb salts of NTO and compared their heat resistance ability with the standard explosives and found in the order: HMX>NTO>KNTO>RDX>PETN>PbNTO>CuNTO. The alkali metal (Li, Na and K) and rare-earth metal (Pr, Nd and Sm) salt hydrates of NTO can also be obtained by adding the corresponding metal hydroxide with aqueous solution of NTO at 40-60^{0}C. The thermal stability of alkali metal salts of NTO has been found to be in the order: Li(NTO).2H_2O>Ca(NTO).H_2O>K(NTO).H_2O. Redman and Spear [69] have prepared Na, K and ammonium salts of NTO as intermediates to prepare a range of heavy metal salts of NTO amely: Ag, Pd, Ba, Cd, Sr, Ni, cupric, mercuric and stannous.

Metal picramates can be easily precipitated by adding a solution of sodium picramate to a solution of salt of the metals. Many workers have studied the thermolysis of picramates of Fe, Co, Ni, Cu, Ag, Zn, Cd, Hg, Ti, Zr and Th. Explosive studies on benzylidene picramates of Fe, Co and Ni, reveal that the thermal stability of these salts increases as the atomic radius of the metal decreases.

Metal salts of picric acid [trinitrophenate (TNP)] may be formed by the action of picric acid on metals, their oxides, carbonates or hydroxides. The properties of picrates of Na, Mg, Al,K, Ca, Cr, Mn, Fe, Co, Ni, Cu, Zn, Sr, Ag, Cd, Ba, Hg, Tl and Pb have been examined by several authors. Di-and mononitrophenols can also form metallic dinitrophenate (DNP) and mononitrophenate (MNP) respectively. Recently, Singh *et al.* [70] have prepared mono- di- and trinitrophenates of Co, Ni and Cu and followed their thermolysis by isothermal and non-isothermal TG. The following mechanism leading to explosion has also been proposed:

$$M[(NO_2)_xC_6H_{5-x}O]_2.yH_2O \xrightarrow[-yH_2O]{<140\ ^0C} M[(NO_2)_xC_6H_{5-x}O]_2 \xrightarrow{>140\ ^0C}$$

$$\{M[C_6H_{5-x}O]_2 + xNO_2\} \xrightarrow[\text{Oxidative decomposition reactions}]{} \text{Explosion} \rightarrow \text{Gaseous products}$$

This involves dehydration below 140 ^{0}C and evolution of NO_2 above 140 ^{0}C followed by oxidative decomposition reactions lead to explosion.

5.9. Conclusion

Nitroguanidine yields several simple products during thermolysis like N_2O, NH_3, CO_2, HCN, urea, cyanogens, cyanoamides, cyanic acid. Besides these simple products it also yields polymeric products like malamin, ammeline, ammelide and cyanogens. NH_2 group act as desensitizing group for aromatic nitro compounds due to its electron donating and/or hydrogen bonding characteristics. Heterocyclic nitro compounds represent explosives of higher performance compared with analogous aromatic system. Metal salts of nitro compounds are used as energetic additives and/or as burning rate modifiers for composite solid rocket propellants.

REFERENCES

[1] H.B. Hass and H. Schechter, *J. Am. Chem. Soc.,* 75 (1953) 1382.
[2] C. S. Choi, *Acta Crystallogr., B37* (1981) 1955.
[3] Y. Oyumi, A. L. Rheingold and T. B. Brill, *Propell. Explos. Pyrot.,* 12 (1987) 46.
[4] K. O. Christe, M. B. Frankel and J. F. Weber, *J. Energ. Mat.* 5 (1987) 83.
[5] J. Stals and M. G. Pitt, *Aust. J. Chem.,* 28 (1975) 2629.
[6] F. Volk, Propell. Explos. Pyrotech., 10 (1985) 139.
[7] A. F. Mc Kay, *Chem. Rev.,* 51 (1952) 301.
[8] Yu Y. Zhong, Su Zhuang, D. Baoru and C.Fubo, *Propell. Explos. Pyrotech.,* 14 (1989) 150.
[9] J. E. Flanagan, M. B. Frankel, J. F. Weber and D. O. Woolery, AFOSR/AFRPL Chemical Rocket Research Meeting, Lancaster, Sept. 1986.
[10] H. B. Hass and E. F. riley *Chem. Rev.,* 32 (1943) 373.
[11] N. Levy and S. D. rose, *Quart. Rev.,* 1 (1948) 358.
[12] N. Kornblum, *Org. reactions,* 12 (1962) 101.
[13] P. Noble Jr., F. G. Borgardt and W. L. Reed, *Chem. Rev.,* 64 (1964) 19.
[14] M. B. Frankel, *tetrahedron,* 19 (1963) 213.
[15] M. J. Kamlet and H. G. Adolph, *J. Org. Chem.,* 33 (1968) 3073.
[16] C. O. Parker, W. D. Emmons, H. A. Rolewicq and K. S. Mc Callum, *Tetrahedron,* 17 (1962) 79.
[17] J. F. Weber and M. B. Frankel, *Propell. Explos. Pyrot.,* 15 (1990) 26.
[18] H. B. Hass, G. B. Bachman and C. C. Addison, *J. Chem. Soc.,* 17 (1952) 935.
[19] F. R. Schenk and G. A. Wetterholm, *US Pat.* 2, 731, 460.
[20] L. Chun Xu, Kogyo Kayaku, 51 (1990) 275.
[21] G. F. Adams and R. W. Shaw, *Annu. Rev. Phys. Chem.,* 43 (1992) 311.
[22] S. Zeman, M. Dimun and S. Truchlik, *Thermochim. Acta,* 78 (1984) 181.
[23] S. Zeman, *Thermochim. Acta,* 31 (1979) 269; 41 (1980) 199; 49 (1981) 219.
[24] T. B. Brill and K. J. James, *Chem. Rev.,* 93 (1993) 2667.
[25] E. E. Gilbert and J. R. Leccacorvi, *Propell. Explos.,* 1 (1976) 89.
[26] J. C. Drummond, *J. Soc., Chem. Ind.,* 41 (1922) 338.
[27] A. T. Nielson, R. L. Atkins and W. P. Norris, *US Pat.* 113 (865) 1981.
[28] A. T. Nielson, R. L. Atkins, W. P. Norris, C. L. Coon and M. E. Sitzmann, *J. Org. Chem.,* 45 (1980) 2341.
[29] C. D. Hutchinson, V. K. Mohan and P. W. Millar, *Propell. Explos. Pyrotech.,* 9 (1984) 161.
[30] W. M. Koppes, H. G. Adolph and M. E. Sitzmann, *US Pat.* 4 (1979) 173, 591.
[31] G. Om Reddy, B. K. M. Murali and A. K. Chatterjee, *Propell. Explos. Pyrotech.,* 8 (1983)29.
[32] W. Naixing, G. Boren and O. Yuxiang, *Propell. Explos. Pyrotech.,* 18 (1992) 111.
[33] A. T. Nielson, S. L. Christian, A. P. Chafin and W. S. Wilson, *J. Org. Chem.,* 51 (1986) 2572.
[34] R. L. Atkins and W. S. Wilson, *J. Org. Chem.,* 51 (1986) 2572.
[35] F. Zeng-guo and C. Boren, *Propell. Explos. Pyrotech.,* 16 (1992) 12.
[36] F. Zeng-guo, C. Boren and L. Zuocai, *Acta Armamentari Sinica,* 1 (1991) 44.

[37] E. P. Eaton, Y. Yiong and R. Gilardi, *J. Am. Chem. Soc.* 115 (1993) 10195.
[38] P. R. Dave, *US Pat.* 5 (1994) 292, 959.
[39] A. P. Marchand, Proc. 4th Adv. *Anal. Detect. Explos. Proc. Intn. Symp.*, (1992) p. 241.
[40] P. N. Preston and C. R. Tennant, *Chem. Rev.*, 72 (1972) 627.
[41] M. Chaykovsky and H. G. Adolph, *J. Energ. Mat.*, 8 (1990) 392.
[42] M. Chaykovsky and H. G. Adolph, *J. Hetrocycl. Chem.*, 28 (1991) 1491.
[43] A. F. Mckay and G. P. Wright, *J. Am. Chem.. Soc.*, 70 (1948) 430.
[44] T. G. Archibald, R. Gilardi, K. Baum and C. George, *J. Org. Chem.* 55 (1990) 2920.
[45] M. A. Hiskey, M. D. Coburn, M. A. Mitchell and B. C. Benicewicz, *J. Heterocycl. Chem.*, 29 (1992) 1855.
[46] G. I. Chipen, R. B. Bokalder and V. Ya Grinshtein, *Khim. Gatero. Soed.*, 2 (1966) 110.
[47] G. Singh, I.P.S. Kapoor, S.K. Tiwari, P.S. Felix, *J. Hazard. Mater.* B 81 (2001) 67.
[48] G. Singh, I.P.S. Kapoor, S.M. Mannan, S.K. Tiwari, *J. Hazard. Mater.* A 68 (1999) 155.
[49] G. Singh, S. Prem Felix, *J. Hazard. Mater.* A90 (2002) 1.
[50] G. Singh, S. Prem Felix, *J. Molec. Struct.* 649 (2003) 71.
[51] M. S. Pevzner, T. N. kulibabina, N. A. Povarova and L. V. Kilina, *Khim. Getero. Soed.*, 8 (1979) 1132.
[52] K. Y. Lee, C. B. Storm, M. A. Hiskey and M. D. Coburn, *J. Energ. Mat.*, 9 (1991) 415.
[53] A. Gao, A. L. Rheingold and T. B. Brill, *Propell. Explos. Pyrotech.* 16 (1991) 97.
[54] G. Singh, C. P. Singh, J. P. Agrawal and A. K. Sikder, *Ind. J. Eng. Mater. Sciences,* 13 (2006) 462.
[55] W. Naixing, C. Boren and Ou Yuxiang, *Propell. Explos. Pyrotech.*, 17 (1992) 265.
[56] T. B. Brill, D. G. Patil, J. Duterque and G. Lengelle, *Combust. Flame,* 95 (1993) 183.
[57] H. Ritter and H. H. Licht, *Propell. Explos. Pyrotech.*, 10 (1985) 147.
[58] T. Oyumi and T. B. Brill, *J. Phys. Chem.*, 91 (1987) 3657.
[59] D. A. Cichra and H. G. *Adolph, Synthesis,* 10 (1983) 830.
[60] S. Zeman and M. Dimun, *CS Pat.* 218, 703,(1986); 227, 907 (1987); 227, 913.
[61] S. Zeman, M. Dimun, V. Kabatova, E. Zemanova, S. Truchlik and B. Belko, *CS Pat.* 238, 442 (1988); 238, 443.
[62] M. Dimun, S. Zeman, S. Truchlik, V. Kabatova and D. Belko, *US Pat.* 238, 440 (1989).
[63] T. B. Brill and C. O. Reese, *J. Phys. Chem.*, 84 (1980) 1.
[64] Y. Oyumi and T. B. Brill, *Combust. Flame,* 62 (1985) 213.
[65] S. F. Palopoli and T. B. Brill, *Combust. Flame,* 87 (1991) 45.
[66] M. B. Frankel and U. D. Woolery, *J. Org. Chem.*, 48 (1983) 611.
[67] K. U. B. Rao, R. R. Soman and H. Singh, *J. Energ. Mat.*, 8 (1990) 99.
[68] X. Yi, H. Rongzu, Y. Chaoquing, F. Cuofu and Z. Jihua, *Propell. Explos. Pyrotech.*, 17 (1992) 298.
[69] L. D. Redman and R. J. Spear, *Report MRL-TN-563,* DODA-AR-006-283, (1989).
[70] G. Singh, M. Yunus and P. Srivastava, *J. Therm. Anal.*, 39 (1993) 135.

Chapter 6

Thermolysis of Transition Metal Nitrate Complexes

Abstract

The preparation, characterization and thermolysis of transition metal nitrate complexes with various ligands have been discussed. The kinetics of decomposition has also been described using model fitting as well as isoconversional method.

6.1. Introduction

The union of a metal cation with one or more neutral molecules or ions (known as ligand) is referred as complex compound. The ligand may be monoatomic or polyatomic and are capable of independent existence in the absence of coordinate bonding to the central atom. Neutral ligand molecules are H_2O, NH_3, ethylenediamine (en), propylenediamine (pn), pyridine (py), CO, NO, etc. and ions are, anions such as Cl^-, Br^-, NO_3^-, SO_4^{2-}, ClO_4^- or cations such as NO^+, $NH_2NH_3^+$, etc. The literature of coordination compounds is so extensive [1] and the number of compounds in this class is very large and diverse having a wide spectrum of applications. Transition metal complexes decompose to respective transition metal oxide (TMOs) and these ultrafine oxides may have interesting electrical, magnetic and catalytic properties [2]. A large number of semiconducting oxides (e.g. CuO, SnO_2, MnO, ZnO, etc) are being explored for application in preparation of environmental sensors for detecting trace level pollutants such as H_2S, CO, NO_x, NH_3, etc. in the atmospheric air [3-8]. When exposed to air, oxygen chemisorbs on the surface of the TMOs to form the species such as O_2^-, O^-, and O^{2-} and influences their surface electrical conductivity. The pollutant gases when present in air react with these species and modulate their conductivity. The change in conductivity is correlated with the pollutant concentration. Hence, by increasing the specific surface area of the TMOs, higher sensitivities can be achieved [9]. Naturally, the nanocrystalline TMOs formed as final decomposition product of the complexes may be better substitute for this application. A good number of pure metal oxides or mixed metal oxides find interesting applications as catalysts in several hydrogenations, dehydrogenation, oxidation and alkylation reactions [10].

The metal complexes consisting of a reducing group (e.g. NH_3, en, pn, etc) as ligand and an oxidizing group (e.g. NO_3^-, ClO_4^-, N_3^-, etc.) as counter anion, exhibit characteristics of energetic compounds. These complexes when raised to sufficient high temperature, whether by direct heating, friction, impact, shock, spark or flame, undergo highly exothermic self-propogative decomposition reaction. The explosive decomposition of such complexes seems to be due to simultaneous presence of both an oxidizing and reducing groups in the same complex molecules. Depending upon the nature of explosivity, these complexes are categorized to lie between those of primary and secondary explosives [11, 12]. On thermolysis, these complexes release the chemical energy available in them and produce gaseous decomposition product and solid residues i.e. respective TMOs [13]. Such complexes may find applications in propellants, explosives and pyrotechnic compositions. Although, TMOs are well known catalysts for burning of composite solid propellants (CSPs) [14-17], many reports are available in open literature which have established that some energetic complexes [18] and salts of NTO [19, 20, 21] are better ballistic additives for CSPs and its components than the TMOs. Both heat release on exothermic decomposition of these compounds and final product of decomposition i.e. TMOs were found to be responsible for enhancement of burning rate. All of these additives were studied to be active during condensed phase thermolysis of CSPs and ammonium perchlorate (AP). Thus, thermolysis studies on transition metal complexes provide systematic guidance for design and development of various devices and chemical formulations based on such complexes.

6.2. Preparation and Characterization:

In general metal nitrate complexes can be prepared following a two step procedure:

I Step – In the first step, the metal carbonates (MCO_3) were treated with 70% nitric acid (HNO_3) to obtain the respective metal nitrates i.e. $M(NO_3)_2$.

$$MCO_3 + 2HNO_3 \rightarrow M(NO_3)_2 + H_2O + CO_2\uparrow$$

The metal nitrates thus obtained were washed with petroleum ether, recrystallised from distilled water and dried over fused $CaCl_2$ in a desiccator.

II Step – Metal nitrate complexes can be prepared by reacting ethanolic solution of various $M(NO_3)_2$ obtained from the first step with ligand following the standard procedures reported in literatures

$$M(NO_3)_2 + \text{ligand} \rightarrow \text{metal nitrate complex}$$

The precipitated complexes of different colors were separated from mother liquor by decantation and washed with methanol and / or recrystallised from aqueous solution and dried. The complexes were characterized by standard gravimetric methods [22], infrared [23, 24] and elemental analysis. Characterisation data of some metal nitrate complexes are presented in table 1.

Table 1. Physical, elemental and spectral parameters of some metal nitrate complexes

Complex	Colour	Elemental analysis (%) Observed (Calculated)				IR ν (cm^{-1})								
		C	H	N	Metal	M-N	$-NH_2$	N-H Str.	N-H Bend	H_2N-CH_2	C-H Str.	C-H Bend	CH_2	NO_3^-
$[Co(eda)_2](NO_3)_2$	Brown	15.3 (15.8)	4.7 (5.7)	26.9 (27.7)	18.9 (19.5)	448	--	--	--	1055	--	--	1470	1365
$[Ni(eda)_2](NO_3)_2$	Violet	14.9 (15.7)	4.5 (5.3)	27.1 (27.7)	18.7 (19.4)	440	--	--	--	1110	--	--	1485	1405
$[Cu(eda)_2](NO_3)_2$	Dark blue	15.1 (15.6)	4.3 (5.2)	26.5 (27.3)	19.8 (20.7)	428	--	--	--	1040	--	--	1455	1350
$[Zn(eda)_2](NO_3)_2$	White	14.8 (15.5)	4.4 (5.2)	26.6 (27.2)	20.1 (21.0)	445	--	--	--	1070	--	--	1460	1395
$[Cu(dien)_2](NO_3)_2$	Dark blue	23.7 (24.4)	5.9 (6.6)	27.6 (28.5)	15.5 (16.1)	428	--	--	--	1035	--	--	1457	1340
$[Zn(dien)_2](NO_3)_2$	White	23.7 (24.3)	5.8 (6.6)	27.5 (28.4)	15.9 (16.5)	440	--	--	--	1110	--	--	1495	1400
$[Cr(pn)_2](NO_3)_2$	Light green	21.8 (22.3)	5.7 (6.2)	25.4 (26.0)	14.8 (15.7)	430	--	--	--	1055	--	--	1460	1365
$[Mn(pn)_2](NO_3)_2$	Brown	21.5 (22.1)	5.7 (6.1)	25.5 (25.8)	15.9 (16.5)	445	--	--	--	1058	--	--	1478	1358
$[Ni(pn)_2](NO_3)_2$	Violet	21.2 (21.8)	5.8 (6.0)	24.5 (25.3)	16.9 (17.8)	438	--	--	--	1108	--	--	1464	1400
$[Cu(pn)_2](NO_3)_2$	Dark blue	20.6 (21.5)	5.7 (6.0)	24.2 (25.0)	18.0 (18.9)	432	--	--	--	1070	--	--	1482	1376
$[Zn(pn)_2](NO_3)_2$	White	20.5 (21.4)	5.3 (5.9)	23.4 (24.9)	18.7 (19.3)	440	--	--	--	1045	--	--	1480	1388
$[Mn(dab)_2](NO_3)_2$	Brown	25.9 (26.3)	5.9 (6.6)	23.1 (22.9)	17.4 (17.8)	494	2980	3244	1578	1068	2886	1410	--	1383
$[Ni(dab)_2](NO_3)_2.2H_2O$	Sky blue	26.8 (26.4)	6.0 (6.6)	24.2 (23.1)	16.8 (17.3)	542	2948	3237	1582	1061	2862	1423	--	1383
$[Cu(dab)_2](NO_3)_2.H_2O$	Blue	25.9 (26.8)	6.1 (6.6)	21.3 (20.5)	17.1 (16.4)	550	2925	3231	1590	1052	2855	1436	--	1383
$[Zn(dab)_2](NO_3)_2$	White	26.6 (27.0)	6.1 (6.7)	22.4 (23.7)	14.9 (15.5)	591	2922	3227	1574	1041	2854	1448	--	1383
$[Ni(dah)_2](NO_3)_2$	Sky blue	35.3 (34.6)	7.5 (7.7)	12.9 (12.5)	14.4 (14.1)	523	2935	3309	1528	1087	2852	1420	--	1383

Table 1. (Continued)

Complex	Colour	Elemental analysis (%) Observed (Calculated)				IR ν (cm^{-1})								
		C	H	N	Metal	M-N	$-NH_2$	N-H Str.	N-H Bend	H_2N-CH_2	C-H Str.	C-H Bend	CH_2	NO_3^-
$[Cu(dah)_2](NO_3)_2$	Blue	34.8 (34.3)	7.9 (7.6)	13.6 (13.3)	15.6 (15.2)	571	2926	3238	1589	1032	2855	1443	--	1383
$[Zn(dah)_2](NO_3)_2$	White	33.5 (34.2)	7.1 (7.5)	13.8 (13.3)	14.9 (15.5)	544	2925	3273	1590	1018	2855	1438	--	1383
$[Co(bipy)_2](NO_3)_2.4H_2O$	Pink	42.9 (42.3)	4.4 (3.70	16.4 (14.8)	9.2 (10.4)	485	--	--	--	--	3066	--	--	1383
$[Ni(bipy)_2](NO_3)_2.2H_2O$	Sky blue	45.4 (44.7)	3.3 (3.7)	14.8 (15.6)	11.3 (10.9)	496	--	--	--	--	3091	--	--	1383
$[Cu(bipy)_2](NO_3)_2$	Blue	46.1 (47.4)	3.9 (3.2)	16.9 (16.6)	11.2 (12.5)	522	--	--	--	--	2990	--	--	1383

Table 2. Some data from thermal studies of nitrate complexes of transition metals with amine ligands

Complex	Step	TG		DTA			DSC		
		Temp. range/^{0}C	%mass loss	Temp range/ ^{0}C	Peak /^{0}C	Nature	Temp range / ^{0}C	Peak / ^{0}C	Nature
$[Cu(en)_2](NO_3)_2$	I	170 – 240	16.5	201 – 241	225	endo			
	II	250 - 265	75.2	242 - 270	248	exo			
$[Co(en)_2](NO_3)_2$	I	178 – 245	17.8	186 – 249	237	endo			
	II	250 - 270	72.4	248 - 272	254	exo			
$[Ni(en)_2](NO_3)_2$	I	180 – 247	17.9	188 – 250	245	endo			
	II	260 - 277	72.8	260 - 280	265	exo			
$[Zn(en)_2](NO_3)_2$	I	182 – 259	18.5	191 – 263	253	endo			
	II	265 - 280	72.9	262 - 285	273	exo			
$[Cr(pn)_2](NO_3)_2$	I	280 – 337	62	300 – 338	321				
	II	340 - 379	16	340 - 388	370				
$[Mn(pn)_2](NO_3)_2$	I	262 – 295	22	258 – 295	282				
	II	297 – 337	23	297-338	315				
	III	395 - 463	33	389-465	424				
$[Ni(pn)_2](NO_3)_2$	I	230-280	43	240-280	254				
	II	335-415	43	350-420	388				
$[Cu(pn)_2](NO_3)_2$	I	220-229	76	219-242	227				
$[Zn(pn)_2](NO_3)_2$	I	230-290	43	244-292	274				
	II	355-405	42	343-425	384				
$[Cu(dien)_2](NO_3)_2$	I	170-183	12.3					180	endo
	II	200-212	25.0					187	endo
	III	248-267	43.0					238	exo
$[Zn(dien)_2](NO_3)_2$	I	240-290	26.2					285	endo
	II	326-400	54.0					314	exo
									Contd..
$[Zn(dab)_2](NO_3)_2$	I	220-280	24.8	260-280	262	endo	266-284	279	
	II	299-305	54.2	299-305	303	exo	309-318	311	endo
									exo
$[Cu(dab)_2](NO_3)_2.H_2O$	I	85-121	4.6	77-117	99	endo	87-115	102	endo
	II	121-248	52.4	--	--	endo	--	--	endo
	III	--	--	195-221	217	exo	206-225	218	exo
	IV	307-470	19.2	315-412	378	exo	320-410	342	exo

Table 2. (Continued)

Complex	Step	TG		DTA			DSC		
		Temp. range/^{0}C	%mass loss	Temp range/ ^{0}C	Peak /^{0}C	Nature	Temp range / ^{0}C	Peak / ^{0}C	Nature
$[Cu(dab)_2](NO_3)_2.2H_2O$	I	82-150	8.7	78-119	103	endo	91-116	102	endo
	II	--	--	215-227	230	endo	221-244	240	endo
	III	200-227	66.5	251-276	268	exo	258-273	266	exo
	IV	315-420	5.1	315-408	374	exo	333-426	395	exo
$[Mn(dab)_2](NO_3)_2$	I	125-216	23.0	148-160	154	endo	158-170	166	endo
	II	218-240	51.2	215-229	221	exo	215-224	220	exo
	III	242-385	4.8	269-379	323	exo	291-389	344	exo
$[Zn(dah)_2](NO_3)_2$	I	--	--	177-217	198	endo	174-214	195	endo
	II	270-302	54.0	271-299	288	exo	267-308	289	exo
	III	440-530	25.0	439-530	485	exo	450-539	496	exo
$[Cu(dah)_2](NO_3)_2$	I	200-255	42	211-252	233	exo	208-248	230	exo
	II	430-510	38	430-509	474	exo	426-504	468	exo
$[Ni(dah)_2](NO_3)_2$	I	200-238	12	202-234	218	endo	208-219	215	endo
	II	230-260	43	236-255	247	exo	236-258	247	exo
	III	350-455	20	358-439	401	exo	354-441	411	exo
$[Co(bipy)_2](NO_3)_2.4H_2O$	I	78-91	9.5	--	96	endo	--	89	endo
	II	155-187	10.1	--	--	endo	--	126	endo
	III	235-258	11.4	--	225	--	--	266	endo
	IV	286-310	42.2	--	307	exo	--	306	exo
$[Ni(bipy)_2](NO_3)_2.2H_2O$	I	125-209	10.0	--	124	endo	--	122	endo
	II	273-316	60.3	--	318	exo	--	326	exo
$[Cu(bipy)_2](NO_3)_2$	I	187-218	32.1	--	213	endo	--		endo
	II	305-324	45.4	--	332	exo	--		exo

Thermal studies on metal complexes have been described by various investigators which are scattered in literature. Thermoanalytical data on various nitrate complexes are compiled in table 2. The effect of various monodentate and bidentate ligands as well as different types of counter anions on thermal stability and decomposition pattern of the complexes are reported. Kinetic analysis have been made based on the data of thermal studies using TG, DTG, DTA, DSC etc. technique and pathways of decomposition have been proposed. Kinetics of thermal decomposition have also been investigated by various investigators.

6.3. Thermolysis of Zn Complexes

Maslowaska *et al* [24] have reported thermal decomposition of $[Zn(en)_2](NO_3)_2.2H_2O$ in the static air using TG and DTA at a heating rate of $5^0C/min$. The complexes were found to decompose in multisteps. All the steps of decomposition gave endothermic peaks in DTA and ZnO was the final product of thermolysis.

Thermal decomposition of various zinc nitrate complexes with ligands such as ethylenediamine (en) [25], propylenediamine (pn) [26], diethylenetriamine (dien) [27], 1,4-diaminobutane (dab) [28] and 1,6-diaminohexane (dah) [29] were studied by Singh *et al.* It has been observed that all these complexes decompose in two steps. First step is endothermic and second one is exothermic in nature (table 2, figures 1-7)

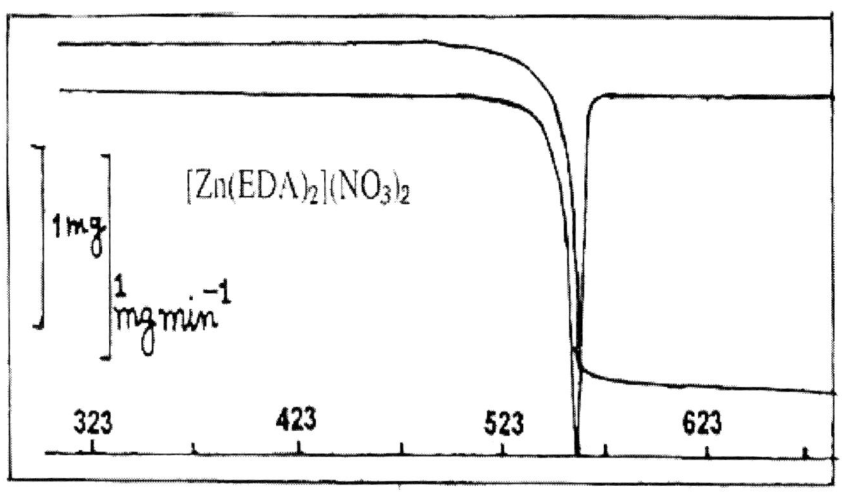

Figure 1. TG-DTG thermograms of $[Zn(EDA)_2](NO_3)_2$.

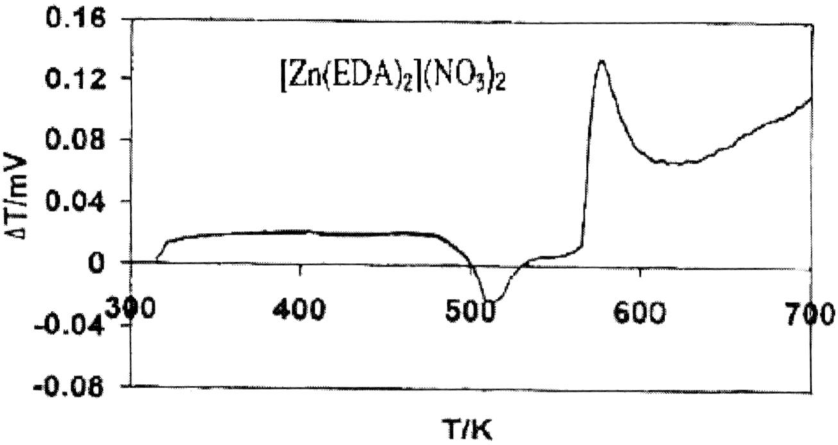

Figure 2. DTA Thermograms of $[Zn(EDA)_2](NO_3)_2$.

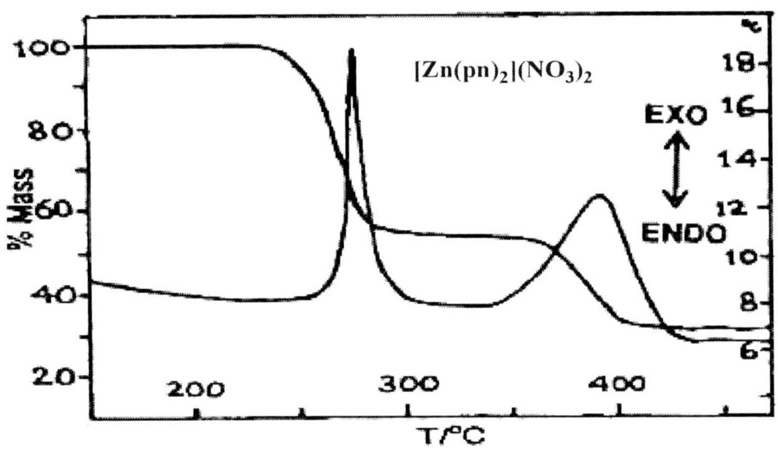

Figure 3. TG-DTA thermograms of $[Zn(pn)_2](NO_3)_2$ in N_2.

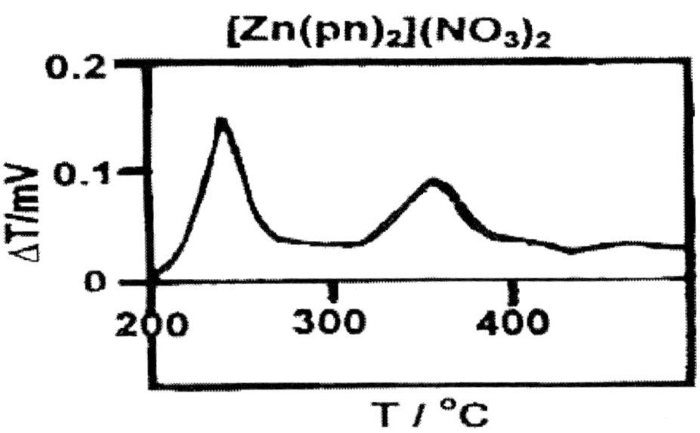

Figure 4. DTA thermogram of $[Zn(pn)_2](NO_3)_2$ obtained in air.

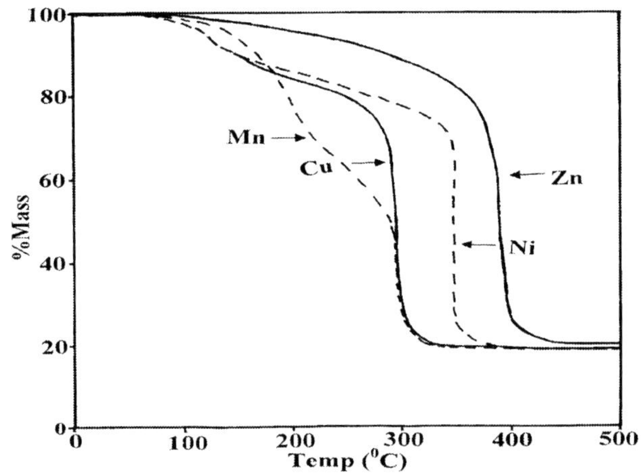

Figure 5. TG curves of dab nitrate complexes in static air atmosphere.

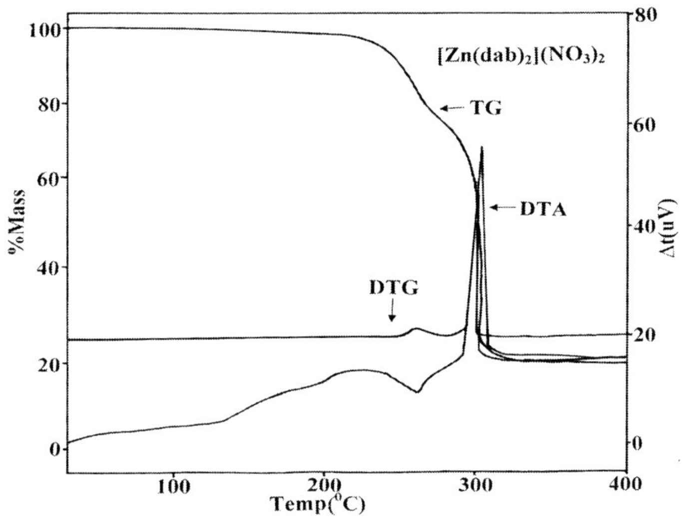

Figure 6. Simultaneous TG-DTG-DTA curves of $[Zn(dab)_2](NO_3)_2$ complex in N_2 atmosphere.

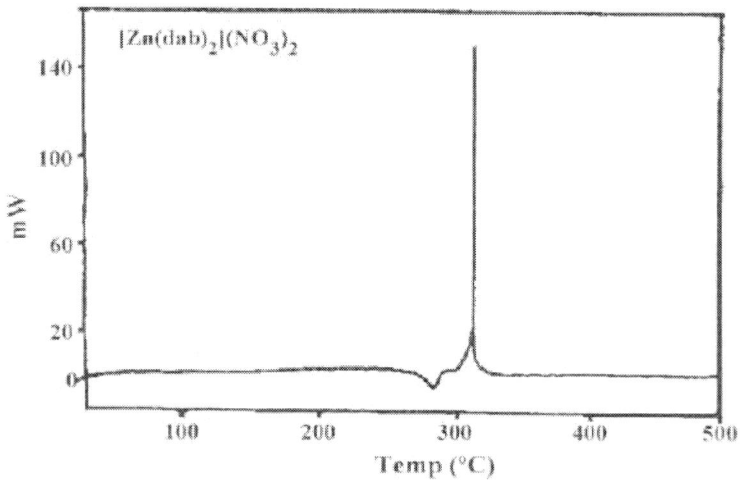

Figure 7. DSC Thermogram of $[Zn(dab)_2](NO_3)_2$.

In general, first step is the partial removal of ligand. After first step mass loss, the left residue ignite exothermically. Final residue correspond to ZnO. After the first step, formation of monoligand complex has been reported. On the basis of TG, DTA and /or DSC data the following mechanism has been proposed for the thermolysis of Zinc nitrate complexes that were found to decompose according to the general scheme 1 given below:

$[Zn(AA)_2](NO_3)_2$	→	$[Zn(AA)](NO_3)_2 + AA$	→	ZnO + gaseous products
(bis(ligand)metal nitrate)		**(mono(ligand)metal nitrate)**		

Scheme 1. Thermal decomposition pathway of zinc nitrate complexes.

An endothermic process immediately followed by an exothermic process has been observed during DTA and/or DSC analysis (figures 1-7). The endothermic process corresponds to the removal of approximately one ligand molecule and exothermic process is due to further decomposition of monoligand complex leading to oxidation-reduction reaction.

A perusal of TG (initial decomposition temperature) and DTA (peak temperature) show that the thermal stability of these complexes were found to be in the order:

dien > pn > dab > en > > dah

In propylenediamine and 1,6-diaminohexane complexes both the step of thermolysis are reported to be exothermic [26, 29]. Just before the first exotherm, an endotherm is present which shows the partial removal / breaking of ligand molecule after which the compound decompose exothermically. In these cases (pn and dah) first exothermic peak is sharp and second is spread over a wide range of temperature (figures 3, 4, table 2) which may be due to the simultaneous decomposition of metal nitrate and oxidation of the metal to metal oxide. In the thermolysis of en, pn, dien and dab complexes, monoligand intermediate, $[M(AA)](NO_3)_2$ have been reported. But in dah complex, there were no indication for the formation of monoligand intermediate. This may be due to the enlarged chain of the ligand molecule so that the carbon chain may break randomly instead of removal of ligand molecule as there is nine membered chelate rings.

Kinetics of thermolysis of these complexes has been studied by using isothermal TG data (figure 8) applying model fitting as well as isoconversional method given by Vyazovkin [30]. The kinetics applying model fitting approach have given average activation energy (E) 108.9, 153.4, 113.9, 76.2 and 105.3 respectively for en, pn, dien, dab and dah complexes.

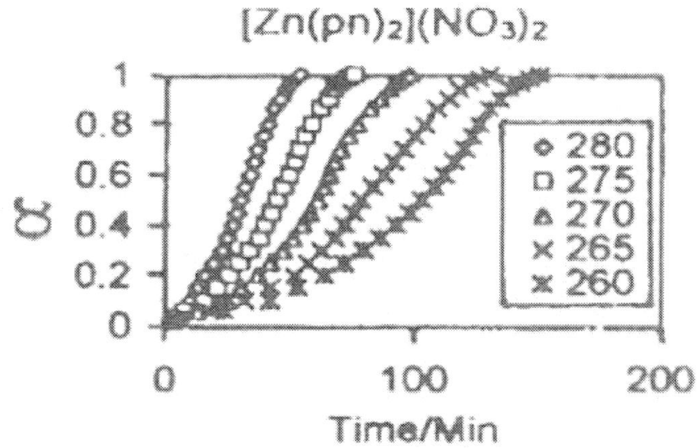

Figure 8. Isothermal TG thermogram of $[Zn(pn)_2](NO_3)_2$.

The percent mass loss (%α) for which the activation energy is calculated is different in different complexes. Due to different extent of conversion (α), a general order of activation energy can not be given and so order of thermal stability can also not be concluded.

Application of isoconversional method to the isothermal TG data have shown that the solid state decomposition of these complexes is not as simple as indicated by the model fitting method. The value of activation energy varies with extent of conversion, α (figure 9)

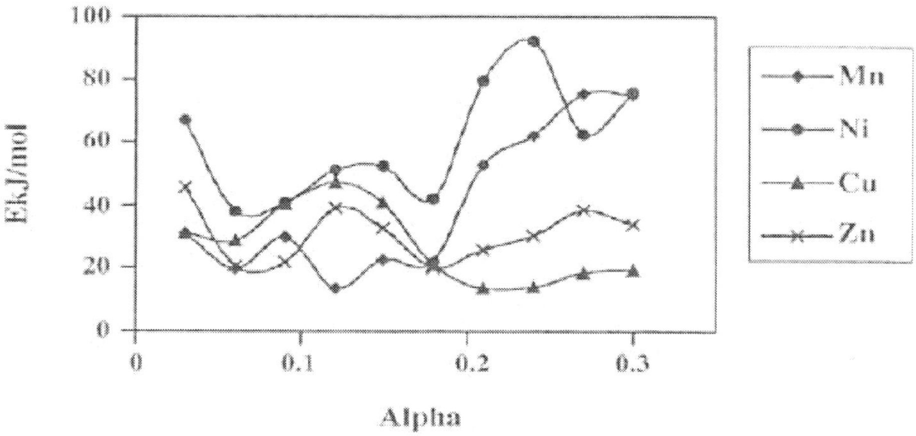

Figure 9. Variation of activation energy with extent of conversion for $[M(dab)_2](NO_3)_2$ complex.

As the solid state thermolysis is so complex, it is not easy to correlate the variation of E value to a particular process. All these complexes are quite stable at room temperature, but they ignite with a noise, light and fumes under sudden high heating. Ignition delay and activation energy of various complexes are summarized in table 3.

Table 3. Ignition delay (D_i) and activation energy of various zinc nitrate complexes

Complex	Ignition delay at 350 ^{0}C in second	Activation energy (E*)
$[Zn(en)_2](NO_3)_2$	85	34.8
$[Zn(pn)_2](NO_3)_2$	93	31.7
$[Zn(dien)_2](NO_3)_2$	125	47.0
$[Zn(dab)_2](NO_3)_2$	1680	131.0
$[Zn(dah)_2](NO_3)_2$	104	33.7

D_i data have shown that the time for ignition at 350 ^{0}C temperature is in the order: dab > dien > dah > pn > en. However, at high temperature the D_i value were found to be changed to that of above order. The activation energy of ignition (table 3) are reported to be in the order: dab > dien > en ~dah > pn. One can predict the thermal stability to be in the same order. Any way, this stability order is different as predicted on the basis of non-isothermal TG analysis (TG, DTA and DSC).

6.4. Thermolysis of Cu Complexes

Thermal studies on numerous copper complexes in regards of various facets have been described by good numbers of investigators, which are scattered in literature. Data on various such complexes, obtained after thermal analyses are compiled in table 2. The effect of various monodentate and bidentate ligands as well as different types of counter anions on thermal stability and decomposition pattern of the complexes are reported. Kinetic analyses have been made based on the data of thermal studies using TG, DTG, DTA, DSC, etc. techniques and pathways of decomposition have been proposed. The general mechanism for thermolysis of Cu complexes may be represented as shown in Scheme 2.

$$CuL_2X_2.nH_2O \xrightarrow[\text{dehydration}]{-xH_2O} Cu\ L2X2 \xrightarrow[\text{deamination}]{-yL} CuX_2 \xrightarrow[\text{deanionation and oxidation}]{-zX+O} CuO$$

Scheme 2. General mechanism of thermolysis of copper complexes.

where, x, y and z represent the appropriate numbers of respective species expelled in various decomposition steps, L = ligands, X = counter anion.

Copper (II) usually has coordination number four with a square coplanar geometry [31] and is known for forming well-defined crystalline complexes of different colours with mono or bidentate ligands. The complexes of Cu (II) with en and tetramethylethylenediamine (tmn) show a number of discontinuities during electrical conductivity measurement in solid state at higher temperatures, indicating the structural phase changes on their heating. In order to understand the changes involved, Prabhumirashi and coworkers [32-34] undertook TG and

DTA on a few Cu and Ni complexes. Recently, they [33, 35] have described the thermal properties of various types of Cu (II) complexes of en and tmn. Their study revealed that the thermolysis of such complexes broadly involves three stages i.e. dehydration (in the case of hydrated ones), deamination and deanionation. The dehydration was a slow process and comprises of broad temperature range while deamination was found to be quite steep ranging over an interval of only a few ^{o}C. The last step consisting of deanionation with simultaneous oxidation of Cu to CuO was also found to be very broad in nature. However, it was found that the three above stated decomposition processes were not isolated but overlap to varying extent. The overlapping of the process thus causes fractional losses of constituent moieties. These complexes undergo phase changes at lower temperatures, which was associated with the absorption of the required latent heat. The phase change was confirmed by endothermic DTA peak, which was not followed by any actual mass loss in accompanying TG. The dehydration was indicated by broad endothermic peak, deamination by sharp exothermic peak and deanionation by broad exothermic peak in DTA respectively. The broad exotherm for the last process was due to simultaneous occurrence of both deanionation and oxidation of Cu to CuO. However, in the case of [Cu (en) $_2$] $(NO_3)_2$ a steep mass loss followed by a quite steep exothermic DTA peak was observed that was attributed to a chain type highly explosive reaction because of simultaneous presence of an oxidizing group (NO_3) and a reducing group (en) in the same complex molecule. These observations are quite comparable to the studies of Wendlandt *et al* [35, 36], Nagase *et al* [37], Srivastava and co-workers [38] and Langfelderova *et al* [39], made on similar complexes. However, in one of the detailed thermal studies (TG and DTA), reported by Wendlandt [40] on the bis–en complex of Cu (II) SO_4 the thermograms appear to differ somewhat from the thermograms as obtained by Prabhumirashi *et al* [34]. This difference may be attributed to the differences in the atmosphere and heating rates employed. Mathew *et al* [31] have reported the detailed thermal studies on bis(ethylenediamine)copper(II)chloride/bromide monohydrates and some other similar complexes containing both monodentate and bidentate ligands in N_2 atmosphere. Unlike the observations of Prabhumirashi *et al*, they have detected the separate mass loss steps for dehydration and deamination. However, in this case also last process i.e. deanionation and volatilization was taking place simultaneously. After dehydration in the first step, one molecule of the en ligand is lost and thus monoligand intermediate is formed [41]. Two endothermic DTA peaks were observed for removal of this first en-molecule, which was due to time lag involved in the breaking of the two coordinate bonds between the bidentate ligand and the metal. It was inferred that in chloride and bromide monohydrate the water molecule occupied in *tetragonal* position, the second being occupied by a halide ion [42, 43]. Hence the thermal decomposition is consistent with the crystal structure of the complex. An intermediate, CuCl was formed at 450 °C and final product was pure Cu at 700 °C, which was confirmed by X-ray powder diffraction [45]. In both the chloride/bromide complexes, the intermediate CuX (X = Cl or Br) is decomposed in final step in which one molecule of en and one chloride/bromide is expelled out [46, 47].

Singh *et al* [25-29, 47] studied the thermolysis of nitrate and perchlorate complexes with the ligands such as ammonia, ethylenediamine (en), peopylenediamine (pn) and diethylenetriiamine (dien) (table 2). Bis(ethylenediamine)copper nitrate complex [25] decompose in two steps (figures 10, 11). An endothermic process immediately followed by an exothermic process has been observed during DTA analysis (figures 11, 12).

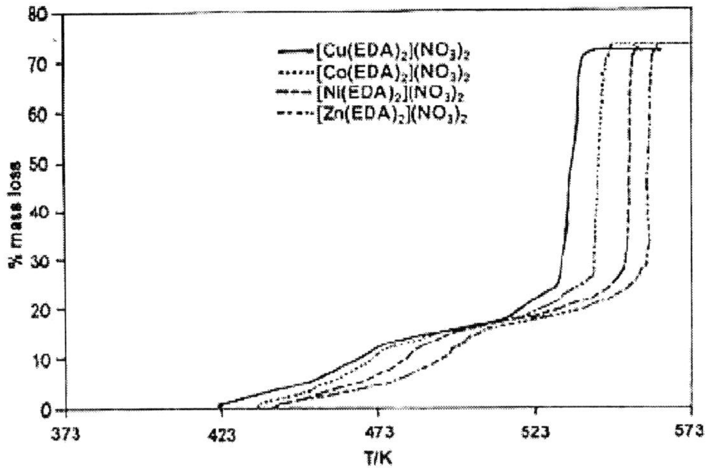

Figure 10. Non-isothermal TG thermograms of BEMN complexes in static air.

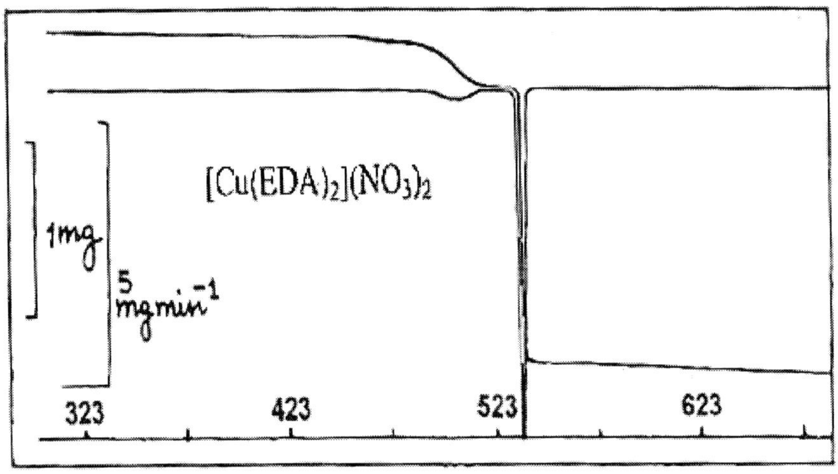

Figure 11. TG-DTG Thermograms of $[Cu(EDA)_2](NO_3)_2$.

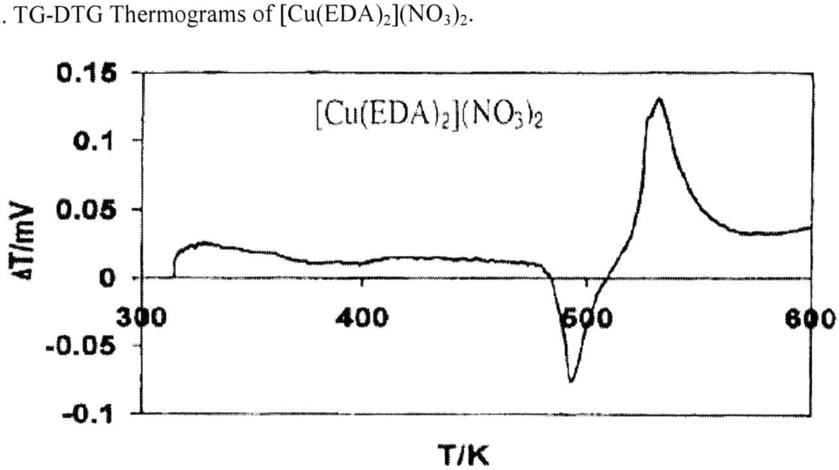

Figure 12. DTA Thermograms of $[Cu(EDA)_2](NO_3)_2$.

The endothermic process corresponds to removal of one en molecule and mono(ethylenediamine) complex is formed. On progressive heating the monoligand complex further decomposed to gaseous products, causing rapid decomposition in its oxidation reduction reaction giving exotherm in DTA. The general scheme of thermolysis has been given below:

$$[Cu(en)_2](NO_3)_2 \xrightarrow{170\text{-}259\ ^0C} [Cu(en)](NO_3)_2 + en$$

$$[Cu(en)](NO_3)_2 \xrightarrow{250\text{-}280\ ^0C} CuO + \text{gaseous products}$$

Scheme 3. Thermolysis pattern of $[Cu(en)_2](NO_3)_2$ complex.

Bis(propylenediamine) copper nitrate complex [26] decompose in one step, which is attributed to a chain type highly explosive reaction (figures 13, 14 and 15) Bis(diethylenetriamine)copper nitrate complex [27] decompose in three steps. The first two stages involves the partial deamination forming monoligand complex. The monoligand species thus formed undergoes final stage of decomposition giving CuO and other gaseous products.

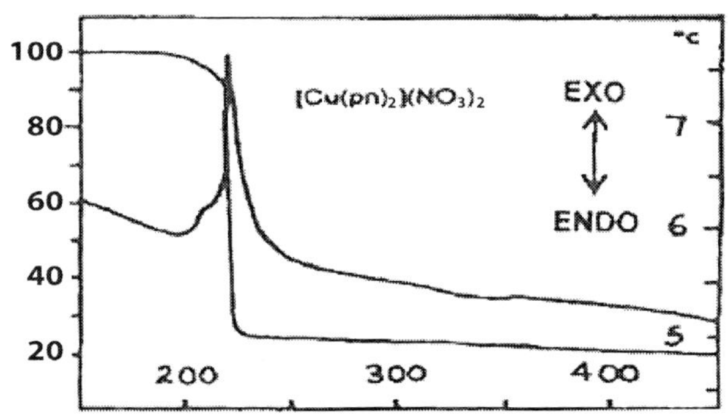

Figure 13. TG-DTA thermograms of $[Cu(pn)_2](NO_3)_2$ in N_2.

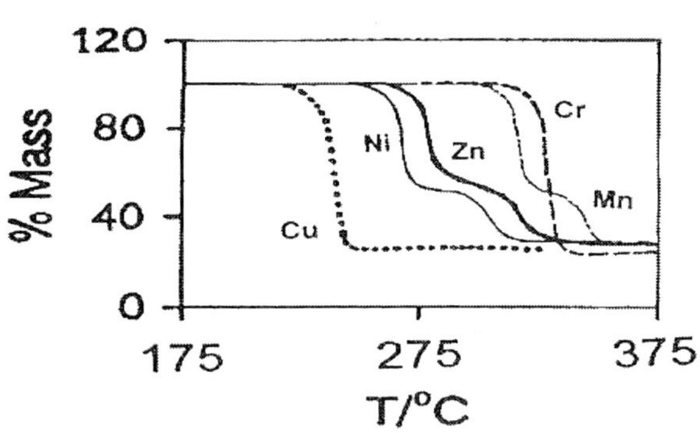

Figure 14. Non-isothermal TG thermograms of complexes recorded in static air.

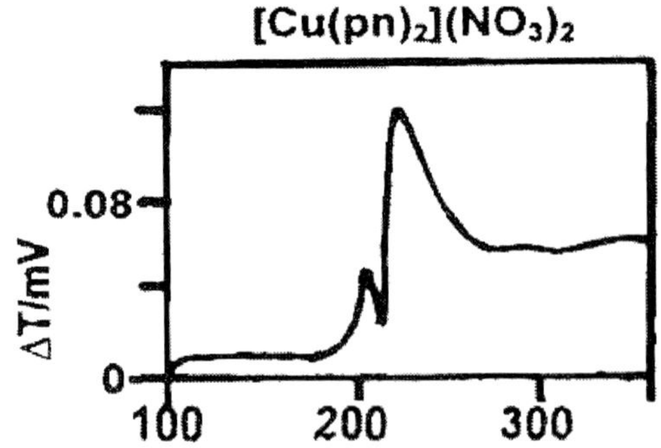

Figure 15. DTA thermogram of $[Cu(pn)_2](NO_3)_2$ obtained in air.

1,4-diaminobutane copper nitrate complex [28] decompose in four steps (table 2, figures 16 and 17). the first step is due to the loss of one H_2O molecule. This step is endothermic. The anhydrous complex $[Cu(dab)_2](NO_3)_2$ decompose and after small mass loss, it ignite at 210^0C giving an exotherm at 218^0C in DTA and DSC. Just prior to this exotherm, an endotherm is shown at ~172^0C which is due to small weight loss prior to sudden weight loss. The small weight loss has been speculated to be the partial removal of ligand molecule. After the ignition of the complex, the remaining residue decomposes slowly exothermically. The final residue was reported to be the CuO.

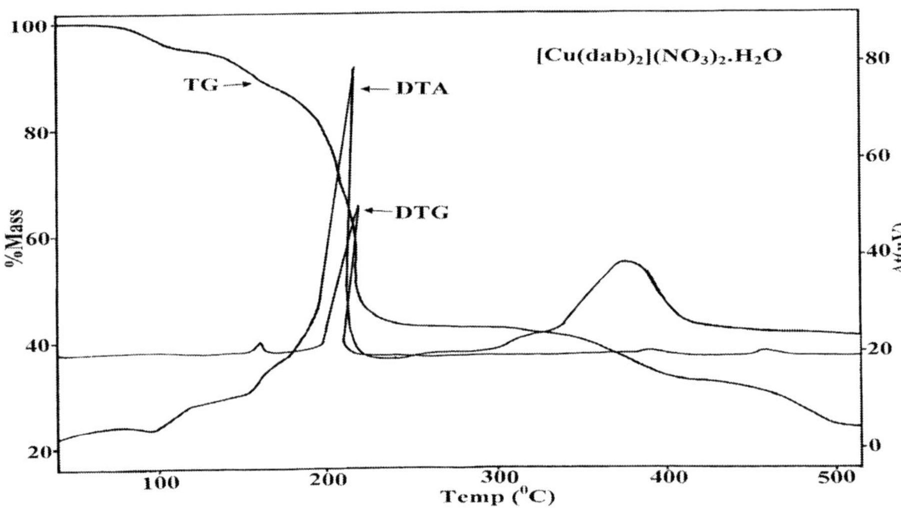

Figure 16. Simultaneous TG-DTG-DTA curves of $[Cu(dab)_2](NO_3)_2$ nitrate complex in N_2 atmosphere.

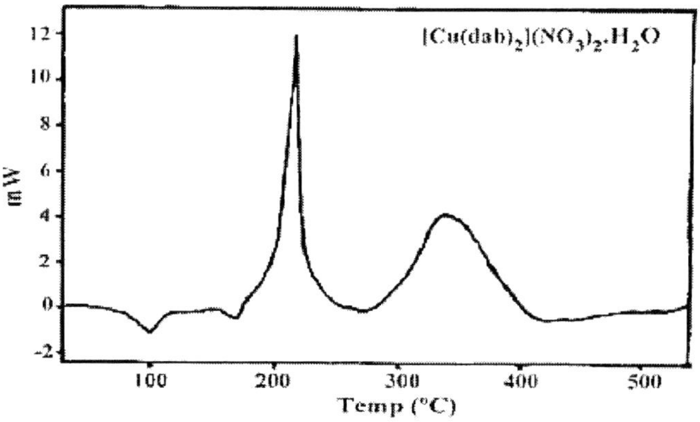

Figure 17. DSC curve of $[Cu(dab)_2](NO_3)_2$ complex.

$[Cu(dah)_2](NO_3)_2$ complex decompose in the same way as $[Zn(dah)_2](NO_3)_2$ complex [29]. $[Cu(bipy)_2](NO_3)_2$ complex [47] decomposes in two steps in both, static air as well as in flowing nitrogen atmosphere and gives two corresponding DTG peaks (figures 18 and 19). In the first step (table 2; 187-218^{0}C), one bipyridine ligand leaves the molecule (~32% wt. loss) giving an endotherm at 213^{0}C in DTA and at 208^{0}C in DSC curve. In second step (305-324^{0}C) the residue $[Cu(bipy)](NO_3)_2$ ignited exothermically and for this weight loss (~45.4%) an exothermic peak at ~332^{0}C was obtained in DTA and DSC curves (figures 19 and 20). Finally, the residues correspond to copper oxide. Thus, the plausible mechanistic pathways for the thermolysis of this complex may be proposed as

$$[Cu(bipy)_2](NO_3)_2 \xrightarrow{187\text{-}218^0C} [Cu(bipy)](NO_3)_2 + bipy$$

$$[Cu(bipy)](NO_3)_2 \xrightarrow{305\text{-}324^0C} CuO + \text{gaseous products}$$

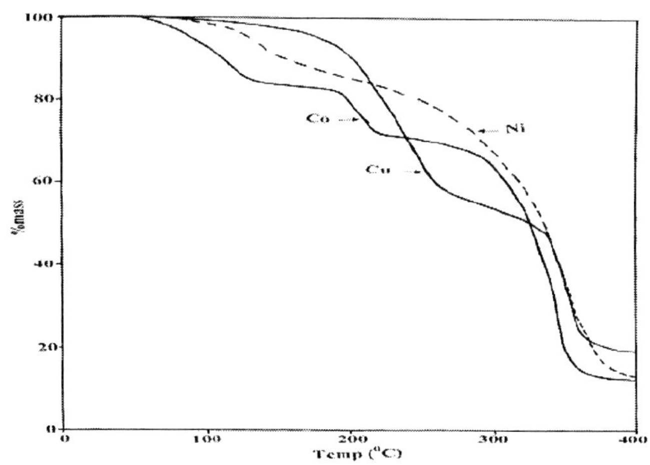

Figure 18. Non-isothermal TGA of the $[Cu(bipy)_2](NO_3)_2$ complex taken in static air atmosphere.

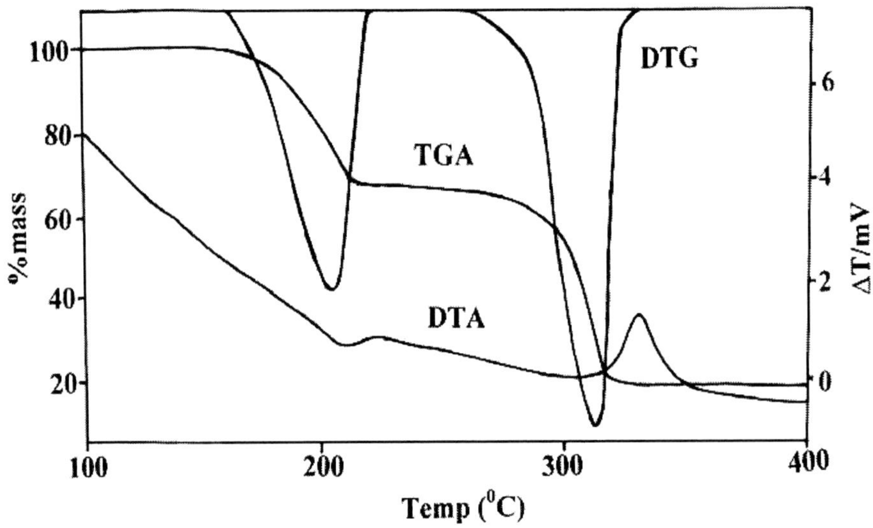

Figure 19. Simultaneous TG-DTG-DTA thermograms of the $[Cu(bipy)_2](NO_3)_2$ complex taken in nitrogen atmosphere.

Table 4. Ignition delay and activation energy of copper nitrate complexes

Complex	Ignition delay at 350^0C(s)	Activation Energy E* (kJ/mol)
$[Cu(eda)_2](NO_3)_2$	55	27.9
$[Cu(pn)_2](NO_3)_2$	90	28.2
$[Cu(dien)_2](NO_3)_2$	92	37.0
$[Cu(dab)_2](NO_3)_2$	103	45.1
$[Cu(dah)_2](NO_3)_2$	100	35.2
$[Cu(bipy)_2](NO_3)_2$	105	27.8

Kinetic analysis from isothermal TG data of these various copper complexes have given the similar results as Zn complexes. Ignition delay and their kinetic analysis have also yielded results similar to that of Zn complex. On the basis of activation energy for ignition delay (table 4), the thermal stability order can be given as:

dab > dien > dah > pn ~ en ~ bipy

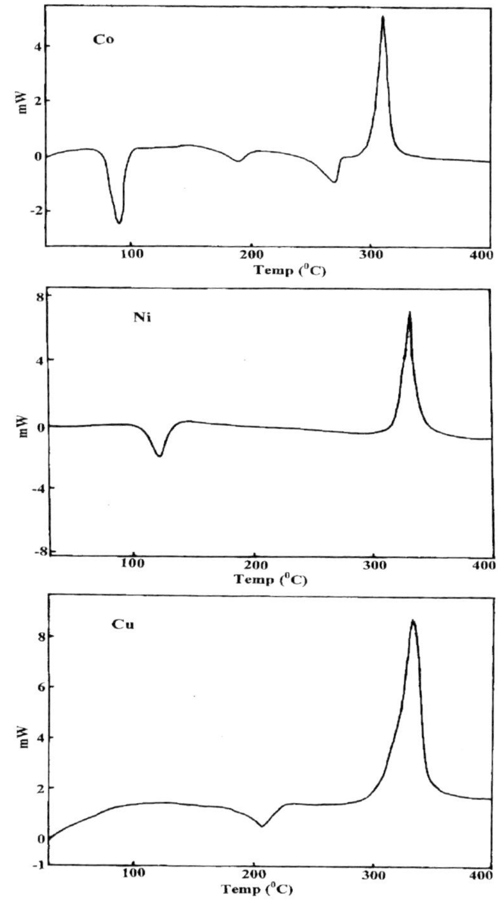

Figure 20. DSC thermograms of the $[Cu(bipy)_2](NO_3)_2$ complexes taken in flowing nitrogen atmosphere.

6.5. THERMOLYSIS OF NI COMPLEXES

Nickel forms well defined crystalline complexes of different colors with various monodentate, polydentate or chelating ligands. $[Ni(NH_3)_6](NO_3)_2$ loss two or three molecules of ammonia prior to an exothermic oxidation reaction (260 ^{0}C) [48]. NH_3 is replaced by NO_3^- during very slow oxidation of Ni^{++} while maintaining the octahedral coordination of the central atom. The complexity of the oxidation reaction involving nitrate decomposition in these coordination is shown by a study which concluded that the analysis of the product of the reaction of $[Ni(aniline)_2](NO_3)_2.4H_2O$. In addition to H_2O and NH_3, some 44 different organic products were detected.

Maslowaska et al [24] studied decomposition of the $[Ni(en)_3](NO_3)_2$ in static air using TG and DTA at a heating rate of 5^{0}C/min. The complexes were found to decompose in multisteps and NiO was obtained as final residue.

Thermolysis of Nickel hydrazinium nitrate complexes [49] have been investigated by Chhabra et al suggest that the nickel complex is relatively less stable than similar cobalt complex. The activation energy determined by DTA and ignition delay measurements correspond to energy of activation of 80±4 kJ/mol fir NiHN and that of 150±8 kJ/mol for CoHN. TG also revealed more rapid decomposition of NiHN than that of CoHN in the temperature range of 215-235^{0}C. High temperature fourier transform-infrared (FT-IR) studied indicated rupture of Ni-N / Co-N bond as the primary step in the thermolysis. DSC of NiHN revealed a violent reaction at 220^{0}C followed by explosion, leading to abnormal curve.

Bis(ethylenediamine)metal nitrate complex [25] decompose in single step similar to Zn complex (figures 21, 22)

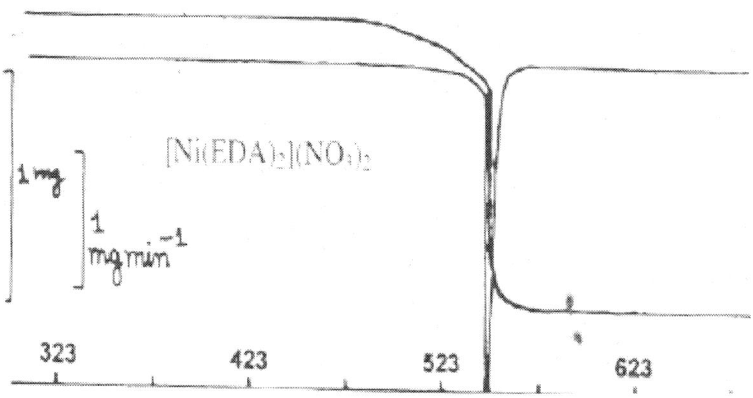

Figure 21. TG-DTG thermograms of $[Ni(EDA)_2](NO_3)_2$.

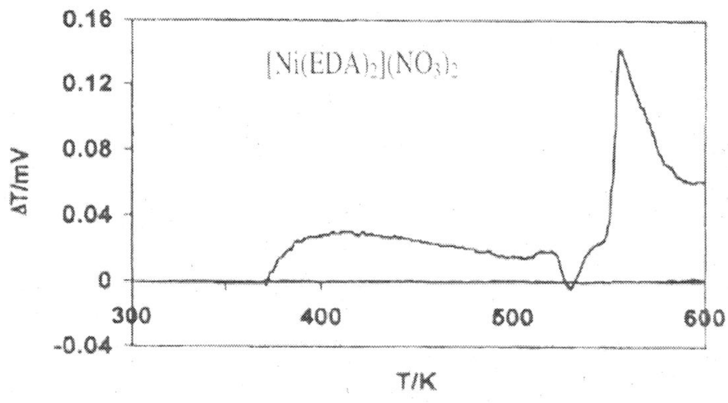

Figure 22. DTA Thermograms of $[Ni(EDA)_2](NO_3)_2$.

Nickel nitrate complexes with propylenediamine [26] decompose in two steps (figures 23, 24). In the first step two ligand molecules are expelled out from the complex and metal nitrate has been formed. The residue decomposes in the second step to their metal oxide. Both the steps are exothermic This is similar to that of Zn complex.

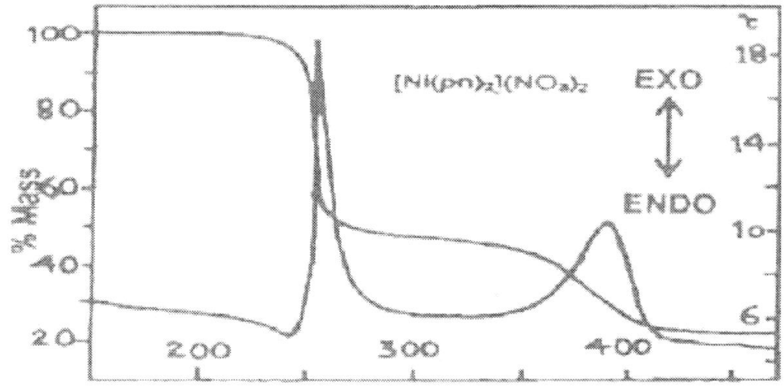

Figure 23. TG-DTA thermograms of $[Ni(pn)_2](NO_3)_2$ in N_2.

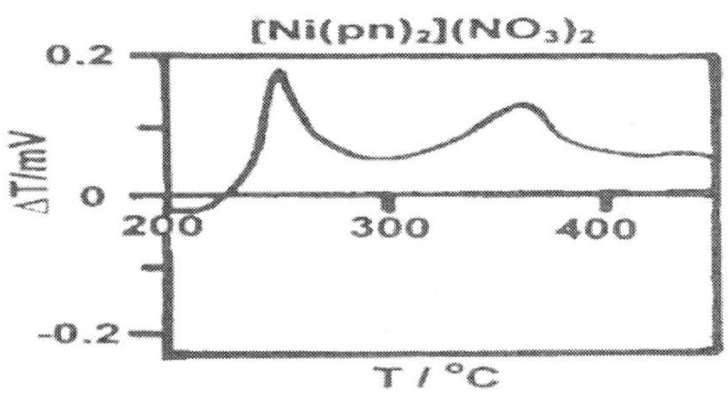

Figure 24. DTA thermogram of $[Ni(pn)_2](NO_3)_2$ obtained in air.

During the decomposition of this complex, monoligand intermediate has not been detected. Nickel nitrate complex with dab ligand decompose similar to that of copper dab nitrate complex. (figures 25 and 26).

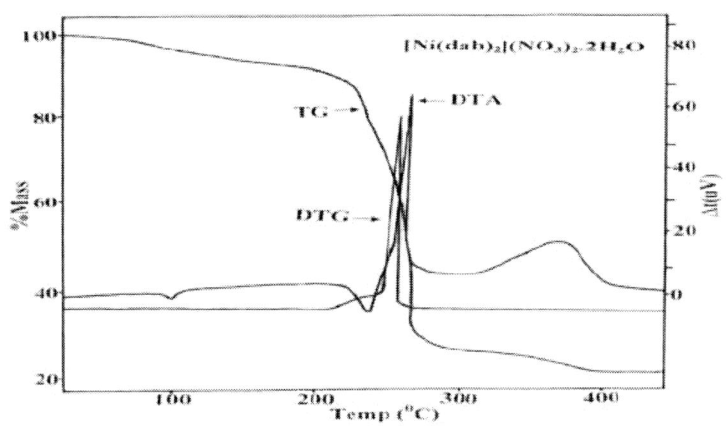

Figure 25. Simultaneous TG-DTG-DTA curves of $[Ni(dab)_2](NO_3)_2$ nitrate complex in N_2 atmosphere.

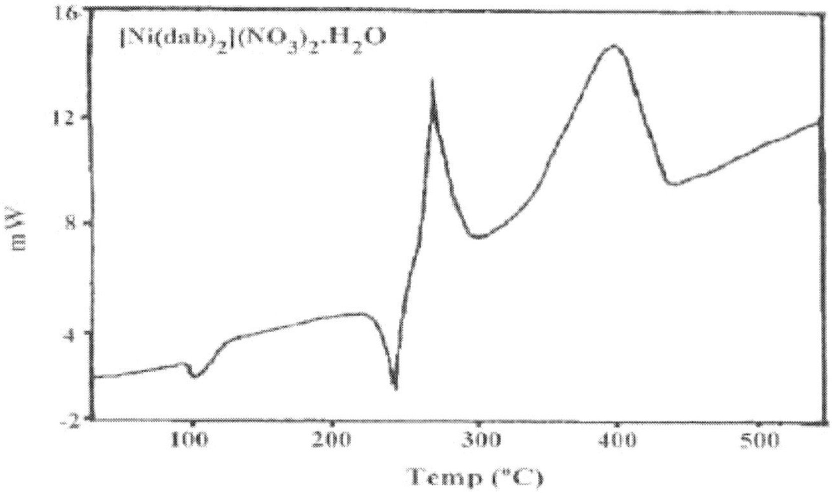

Figure 26. DSC curve of $[Ni(dab)_2](NO_3)_2$ complex.

Nickel bipyridine nitrate complex [47] decomposed in two steps (figures 18 and 27) The first step is the loss of two water molecules giving an endotherm in DTA and DSC (figures 20 and 27). The dehydrated complex decomposes giving nickel oxide as residue and an exotherm. The plausible mechanism for the thermolysis has been proposed as:

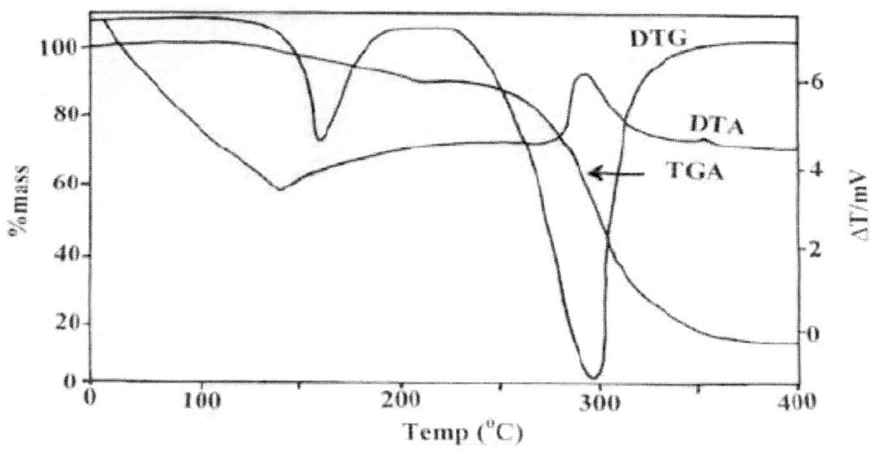

Figure 27. Simultaneous TG-DTG-DTA thermograms of the nickel complex taken in nitrogen atmosphere.

$$[Ni(bipy)_2](NO_3)_2.2H_2O \xrightarrow{125\text{-}209\ ^0C} [Ni(bipy)_2](NO_3)_2 + 2H_2O$$

$$[Ni(bipy)_2](NO_3)_2 \xrightarrow{273\text{-}320\ ^0C} NiO + \text{gaseous products}$$

Nickel nitrate complex with dah ligand [29] have similar thermolytic pattern as Cu and Zn complex. Kinetic analysis from isothermal TG data of nickel complexes have also yield similar results as in case of Zn and copper complexes. Using starting decomposition temperature (table 2) the thermal stability order of the Ni complex can be given as:

pn > dah > en > bipy > dab

Ignition delay stidies of nickel complex has also been reported (table 5)

Table 5. Ignition delay and their activation energy of nickel nitrate complexes

Complex	Ignition delay (s) at 375^0C	Activation energy (kJ/mol)
$[Ni(en)_2](NO_3)_2$	94	34.5
$[Ni(pn)_2](NO_3)_2$	93	30.5
$[Ni(dab)_2](NO_3)_2$	108	52.5
$[Ni(dah)_2](NO_3)_2$	102	34.0
$[Ni(bipy)_2](NO_3)_2.2H_2O$	123	36.8

On the basis of ignition delay data the thermal stability order can be given as:

bipy > dab > dah > en ~pn

6.6. Thermolysis of Cobalt Complexes

Chang and Wendlandt [50-52] investigated several series of complexes of the type (a) cis- and trans- $[Co(en)_2(H_2O)_2](NO_3)_2$ + NH_4X and (b) cis- and trans- $[Co(NH_3)_4(H_2O)_2](NO_3)_3$ + NH_4X. Maslowska and Szmich [24] have carried out TG and DTA on Co (II) complexes of thiosalicylic acid (SR) and ethylenediamine (en) like, [Co(SR) and $[Co(en)_2](NO_3)_2$ in static air atmosphere at a heating rate of 5 °C min^{-1}. They found that both the complexes decompose in multistep and cause endo effect in DTA. Finally, CoO was obtained as decomposition product. Wendlandt [53] shown that thermal stabilities of salts containing the $[Co(NH_3)_6]^{3+}$ ion (heated in helium at 5 °C min^{-1}) vary with the anion present (Cl^-, Br^-, NO_3^- and SO_4^{2-}) and that these anions also influence the chemical changes which occur. $[Co(NH_3)_6](NO_3)_3$ was reported to decompose exothermically in single step and gaseous products of decomposition in order of abundance: H_2O, N_2, NH_3 and N_2O. Wendlandt and Smith [54] also studied the reaction of pentamminehalocobalt (III) nitrates, $[Co(NH_3)_5X](NO_3)_2$ using TG between 127 and 177 °C. They found that the stability sequence for the initial dissociation step was (X =) $F^- = Cl^- > Br^- > I^-$.

Thermolysis and kinetics of decomposition of cobalt nitrate complex with en and 4,4`-bipyridine ligand have been iestigated by Singh et al [25, 47] (tables 1 and 2). Cobalt nitrate complex with en ligand decompose in two steps (figures 28 and 29).

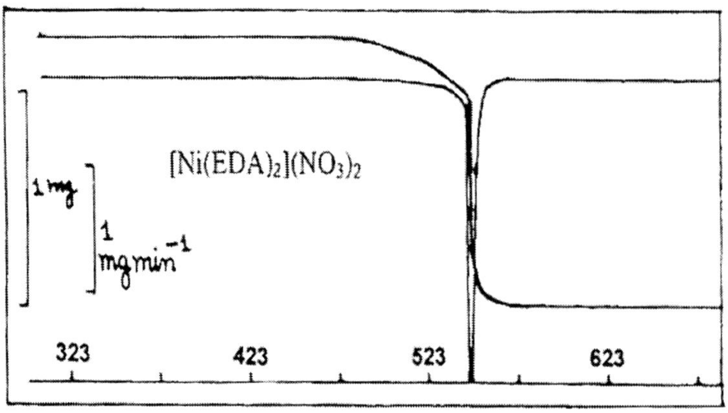

Figure 28. TG-DTG thermograms of $[Co(EDA)_2](NO_3)_2$.

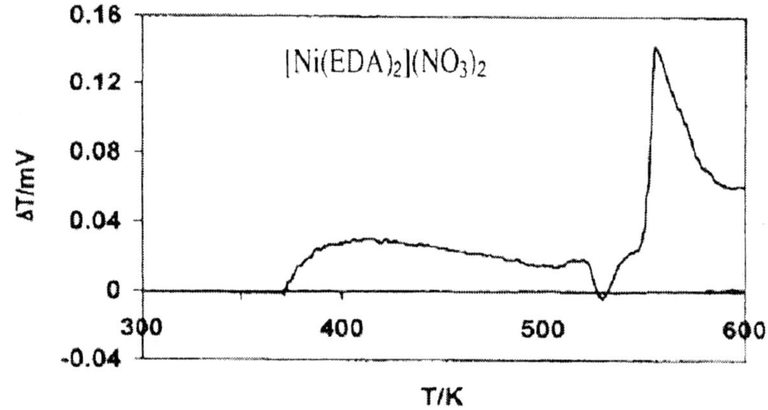

Figure 29. DTA Thermograms of $[Co(EDA)_2](NO_3)_2$.

The various kinetic parameters of thermolysis are summarized in table 2. The pathway of decomposition are also similar to the respective copper complexes.

Thermolysis behaviuor of coordination compound $[Co(urea)_6](NO_3)_2$ have been studied by Carp et al [55]. The coarse of thermal decomposition of this complex starts with two phase transitions (melting and Oh to Td configuration) and continues with seven mass loss steps. According to thermogravimetric and magnetic investigations a dimeric compound $[Co(biuret)(NCO)_2(NO_3)_2$ has been assumed to arise. A gradual oxidation of Co^{++} to Co^{+++} starts at ~250^0C and progresses till 550^0C, when the last nitrate ion undergoes decomposition. At 550^0C, Co_3O_4 with mean crystalline size of ~150^0C has been identified.

Cobalt bipyridine nitrate complex $[Co(bipy)_2](NO_3)_2.4H_2O$ was found to be polymeric as identified by crystal structure (figure 30). TG curve (figures 18 and 31) of this complex shows that it decompose in four steps in static air as well as in flowing nitrogen atmosphere. In the very first step four water molecules leave out. The middle two steps are due to removal of one bipyridine molecule. In fourth step, residue ignite with light and a sharp weight loss is observed which has been speculated due o the exothermic decomposition of $[Co(bipy)](NO_3)_2$ residue. Finally a residue corresponding to cobalt oxide is left. The following mechanism of thermolysis has been proposed:

$$[Co(bipy)_2](NO_3)_2.4H_2O \xrightarrow{95\,^0C} [Co(bipy)_2](NO_3)_2 + 4H_2O$$

$$[Co(bipy)_2]\,(NO_3)_2 \xrightarrow{190\text{-}312^0C} [Co(bipy)](NO_3)_2 + bipy$$

$$[Co(bipy)](NO_3)_2 \xrightarrow{320\,^0C} CoO + \text{gaseous products}$$

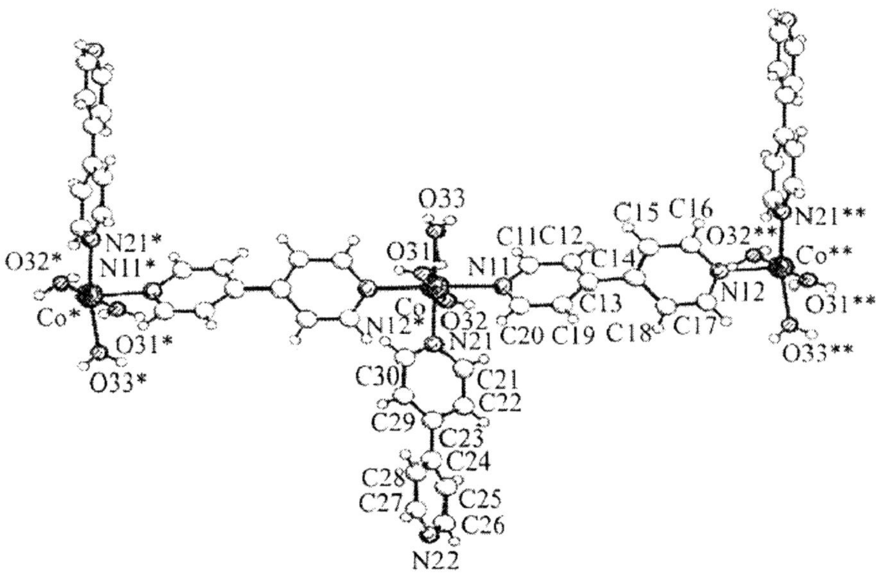

Figure 30. Crystal structure of cobalt bipyridine complex.

First two steps are endothermic whereas fourth step was reported to be exothermic (figure 31).

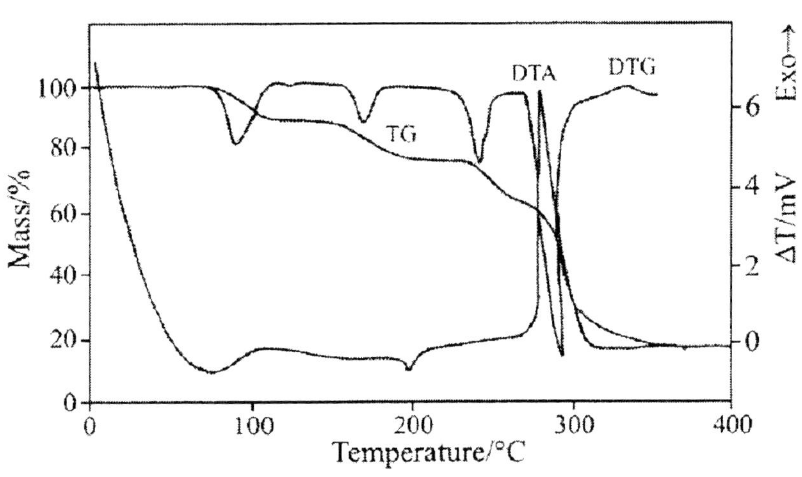

Figure 31. Simultaneous TG-DTG-DTA curve of cobalt bipyridine complex.

$[Co(en)_2](NO_3)_2$ complex reported to decompose in same way as similar Ni and Zn complexes. Value of ignition delay and activation energy (table 6) shows that bipyridine complex is more stable than en complex.

Table 6. Ignition delay and activation energy of cobalt nitrate complexes

Complex	Ignition delay (s) at temperature				E (kJ/mol)	E* (kJ/mol)
	375^0C	400^0C	425^0C	450^0C		
$[Co(eda)_2](NO_3)_2$	53.3	43.3	38.3	33.7	94.4	30.6
$[Co(bipy)_2](NO_3)_2.4H_2O$	74	55	48	42	45.6	36.8

E = activation energy of thermolysis as temperature increases
E* = activation energy of ignition.

Evaluation of kinetics of thermolysis have given same information as Zn, Cu and Ni complexes.

6.7. Thermolysis of Mn Nitrate Complexes

Thermolysis of $[Mn(pn)_2](NO_3)_2$ [26] and $[Mn(dab)_2](NO_3)_2$ [28]has been reported. Deamination of the first complex takes place in two steps (figures 32 and 33) which may be due to higher thermal stability of mono(propylenediamine)manganese nitrate formed during the thermolysis of the pure complex. Both the steps are exothermic.

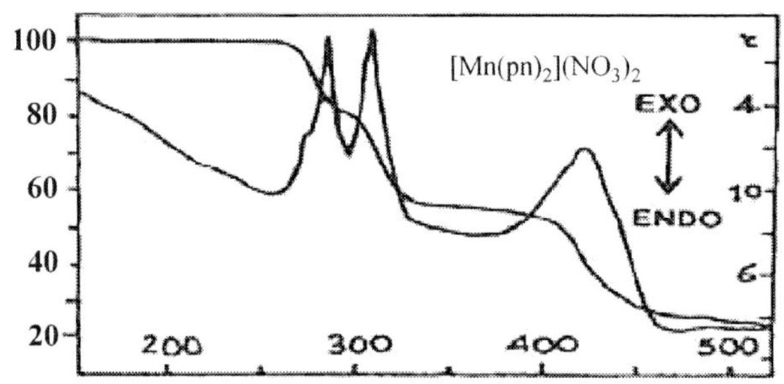

Figure 32. TG-DTA thermograms of $[Mn(pn)_2](NO_3)_2$ in N_2

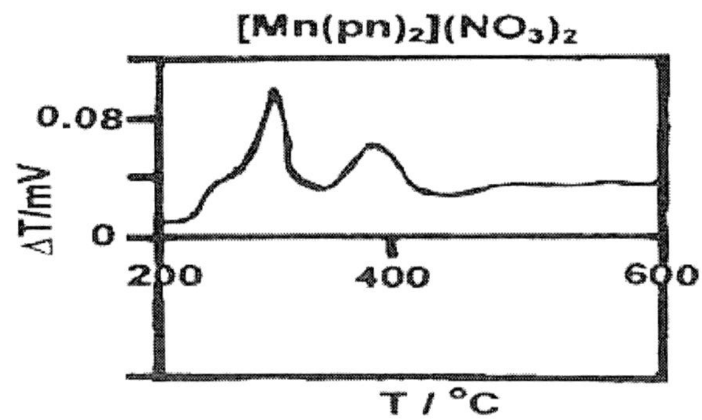

Figure 33. DTA thermogram of $[Mn(pn)_2](NO_3)_2$ obtained in air.

$[Mn(dab)_2](NO_3)_2$ complex decomposes in three steps (figures 34 and 35). In the first step one ligand leave off and is endothermic. In second step, the moiety $[Mn(dab)](NO_3)_2$ ignite and gives a sharp weight loss producing an exotherm in DTA and DSC. The ignited residue even decompose slowly exothermically. Final residue corresponds to manganese oxide.

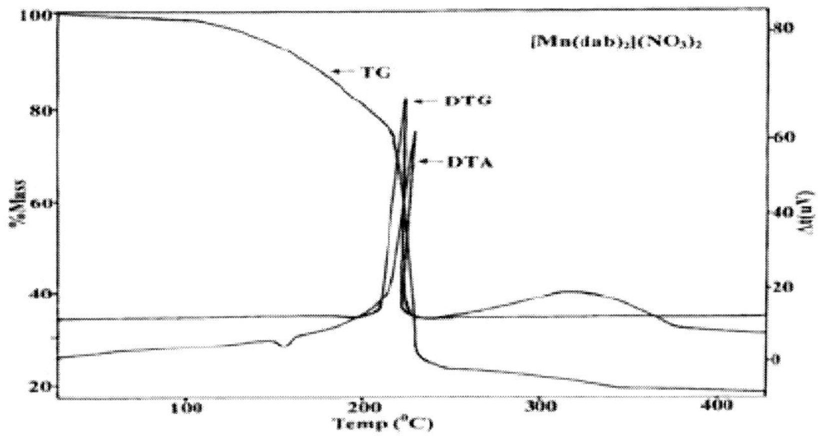

Figure 34. Simultaneous TG-DTG-DTA curves of $[Mn(dab)_2](NO_3)_2$ nitrate complex in N_2 atmosphere.

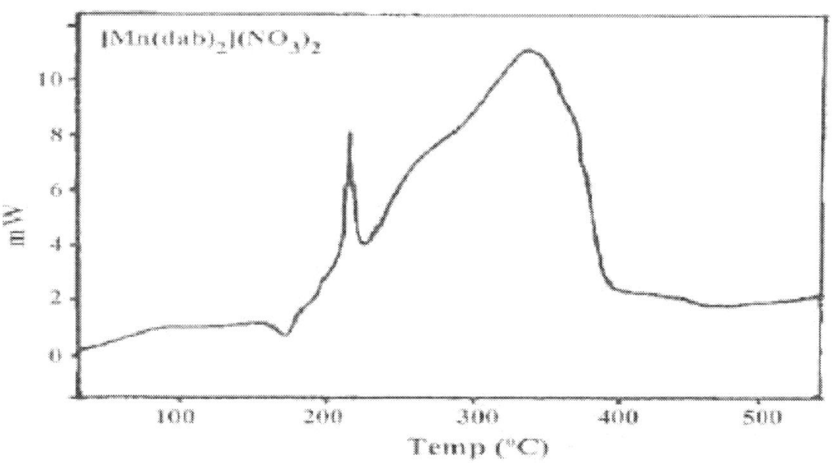

Figure 35. DSC curve of $[Mn(dab)_2](NO_3)_2$ complex.

Table 7. Ignition delay and activation energy of manganese nitrate complexes

Complex	D_i at 350^0C (s)	E* (kJ/mol)	E (kJ/mol)
$[Mn(dab)_2](NO_3)_2$	65	40.7	88.8
$[Mn(pn)_2](NO_3)_2$	97	32.8	154.2

Owing to the value of ignition delay and activation energy of ignition shows the greater thermal stability of pn complex over dab complex. Kinetic analyses have shown similar results as in case of Zn, Cu, Ni and Co nitrate complexes.

6.8. Conclusion

All the metal nitrate complexes with various amine ligands decompose in multisteps. First step is slow and endothermic in nature due to partial removal of ligand. Further steps are rapid and exothermic. In oxidative air atmosphere rate of decomposition is fast in comparison to inert atmosphere. Application of model fitting for the kinetic analysis from isothermal data, fails to explain the complexity of thermolysis as it gives single value of activation energy which can not be assigned to a particular process. On the other hand isoconversional method describes well the complexity of even single step decomposition that yields a series of E values as a function of extent of conversion, α. The ignition delay and E* have been found to be maximum for Zn complexes for a given ligand.

References

[1] G. Wilkinson, R.D. Gillard and J.A. Mc Cleverty (Eds.*), Comprehensive Coordination Chemistry,* Pergamon, Oxford, 1987, Vols 1 – 7.

[2] S.Y. Sawant, K.R. Kannan and V.M.S. Verenkar, in: *Proceedings of Thirteenth National Symposium on Thermal Analysis, Thermans* 2002, pp 154.

[3] P.T. Moseley and D.E. Williams, *Sensors and Actuators B,* 1 (1990) 113.

[4] N. Yamazoe and N. Miura, in: *Chemical Sensors Technology,* Vol. 4 (Ed. S. Yamaguchi) Elsevier, 1992, 19.

[5] D.E. Williams, *Anal. Proc.,* 28 (1991) 366.\

[6] G. Mangamma, V. Jayaraman, T. Gnanasekaran and G. Periaswami, *Sensors and Actuators B,* 53 (1999) 133.

[7] V. Jayaraman, K.I. Gnanaseker, E. Prabhu, T. Gnanasekaran and G. Periaswami, *Sensors and Actuators B,* 54 (1999) 147.

[8] V. Jayaraman, K.I. Gnanaseker, E. Prabhu, T. Gnanasekaran and G. Periaswami, *Sensors and Actuators B,* 54 (1999) 175.

[9] R. Sridharan, V. Jayaraman, R. Ganesan and T. Gnanasekaran, in: *Proceedings of Thirteenth National Symposium on Thermal Analysis,* Thermans 2002, pp 15.

[10] R. Prasad, in: Proceedings of Thirteenth National Symposium on Thermal Analysis, Thermans 2002, pp 106.

[11] W. Friendrich and P. Vervoost, CA 21 (9124) 1184.

[12] K.C. Patil and V.R. Pai Verneker, *Combust. Flame.,* 25 (1978) 387.

[13] G. Singh, I.P.S. Kapoor, S.M. Mannan and J. Kaur, J. *Hazard. Mater., A* 79 (2000) 1.

[14] H.H. Kung, Transition Metal Oxides, *Surface Chemistry and Catalysts,* Vol. 45, Elsevier, Amsterdam, 1998.

[15] Shin – Ming Shen, Sun – I – Chen and Bur – Horng Wu, *Thremochim. Acta,* 223 (1993) 135.

[16] F. Solymosi, Structure and Stability of Salts of Halogen Oxiacids in the Solid Phase, John Wiley and Sons, New York, 1977, pp 195.

[17] P.W.M. Jacobs and H.M. Whitehead, *Chem. Rev.,* 69 (1969) 551.

[18] K.C. Patil and V.R. Pai Verneker, *Combust. Flame,* 25 (1975) 387.

[19] G. Singh, I.P.S. Kapoor, S.K. Tiwari, S.P. Felix, K.N. Ninan and T.L. Varghese, *J. Energ. Mater.*, 20/3 (2003) 309.
[20] G. Singh and S.P. Felix, *Combust. Flame,* 132 (2003) 153.
[21] A. I. Vogel, Text Book of Quantitative Inoragnic Analysis, J. Bassett, R. C. Denny, G. H. Jeffery, J. Mendham, (4th Ed) Longman, London, 1985.
[22] K. Nakamoto, Infrared and Raman Spectra of Inorganic and Co-ordination Compounds, Wiley and Sons, New York, 1978, p. 206.
[23] C. W. Robert, Handbook of Physics and Chemistry, Vol. 66, CRC Press, Florida, 1996. p. 197.
[24] J. Maslowska and J. Szmich, *J. Therm. Anal,* 36 (1990) 2701.
[25] Gurdip Singh and Durgesh Kumar Pandey, *Propell. Explo. Pyrot.* 28 (2003) 5.
[26] G. Singh and D. K. Pandey, *Combust. Flame.,* 135 (2003) 135.
[27] G. Singh and D. K. Pandey, *J. Therm. Anal. Calorimetry,* 76 (2004) 507.
[28] Gurdip Singh, Chandra Prakash Singh and S. M. Mannan, *J. Hazarad. Mater.,* b122 (2005) 111.
[29] Gurdip Singh, Chandra Prakash Singh and S. M. Mannan, *J. Hazarad. Mater.,* A135 (2006) 10.
[30] S. Vyazovkin, C. A. Wight, *J. Phys. Chem A,* 101 (1997) 8279.
[31] S. Mathew, C.G.R. Nair and K.N. Ninan, *Thermochim. Acta,* 181 (1991) 253.
[32] L.S. Prabhumirashi, G.N. Natu and J.K. Khoje, *J. Therm. Anal.,* 35 (1989) 1097.
[33] L.S. Prabhumirashi, G.N. Natu and S.R. Nayak, *J. Therm. Anal.,* 35 (1989) 1105.
[34] L.S. Prabhumirashi and J.K. Khoje, *Thermochim. Acta,* 383 (2002) 109.
[35] T.D. George and W.W. Wendlandt, *J. Inorg. Nucl. Chem.,* 25 (1963) 395.
[36] W.W. Wendlandt, The Thermal Properties of Transition Metal Amine Complexes, Elsevier, Amsterdam, 1967 (Chapter 2).
[37] K. Nagase and H. Yokobayash, *Bull. Chem. Soc. Jpn.,* 47 (1974) 2036.
[38] P.C. Srivastava, B.N. Singh, S.D. Adhya and K.C. Banerji, *J. Therm. Anal.,* 27 (1983) 263.
[39] H. Langfelderova, V. Karla, M. Lirkesova and J. Gazo, *J. Therm. Anal.,* 26 (1983) 95.
[40] W.W. Wendlandt, *J. Inorg. Nucl. Chem.,* 25 (1963) 833.
[41] D. G. Patil, S. R. Jain and T. B. Brill, *Propellants. Explos. Pyrotech.*, 17 (1992) 99.
[42] F. Mazzi, *Rand. Soc. Mineral. Ital.,* 9 (1953) 148.
[43] R.D. Ball, D. Hall, C.E.F. Pickard and T.N. Waters, *J. Chem. Soc. A,* 1967, 1435.
[44] Hannewalt Method Search Manual, International Centre for Diffraction Data, *JCPDS,* Swarthmore, PA pp 748.
[45] J.E. House, Jr. and F.M. Tahir, *Thermochim. Acta*, 117 (1987) 371.
[46] J.R. Allan, B.R. Carson, D.L. Gerrard and J. Birnie, *Thermochim. Acta*, 160 (1990) 329.
[47] G. Singh, C. P. Singh and R. Frohlich, *J. Therm anal. Cal.,* 85 (2006) 2, 425.
[48] M. Prost, E. Garbowski and M.V. Mathieu, *J. Chim. Phys.,* 71 (1974) 1519.
[49] J. S. Chhabra, M. B. Talwar, P. S. Makashir, S. N. Asthana and Haridwar Singh, *J. Hazard. Mater.,* A99 (2003) 225.
[50] F.C. Chang and W.W. Wendlandt, *J. Inorg. Nucl. Chem.,* 32 (1970) 3535.
[51] F.C. Chang and W.W. Wendlandt, *Thermochim. Acta,* 2 (1970) 293.
[52] F.C. Chang and W.W. Wendlandt, *Thermochim. Acta*, 3 (1971) 69.
[53] W.W. Wendlandt, *J. Inorg. Nucl. Chem.,* 25 (1963) 545.

[54] W.W. Wendlandt and J.P. Smith, *J. Inorg. Nucl. Chem.,* 25 (1963) 843,1267; 26 (1964) 445.
[55] O. Carp, L. Patron and A. Reller, *J. Therm. Anal. Cal.* 73 (2003) 867.

Chapter 7

Thermolysis of 5-Nitro-2,4-Dihydro-3H-1,2,4-Triazole-3-One (NTO) and Its Salts

Abstract

NTO is an insensitive high explosive compound having reasonably good performance parameters as an explosive. Owing to the ease of formation of salts, a large number of metal and amine salts of NTO are also known. Considerable research has been done on the thermolysis of NTO and its salts so that to understand the decomposition pathways and to evaluate kinetic parameters. Different research groups employed various spectroscopic, thermochemical, photothermal and chromatographic techniques to study the decomposition of NTO. It is now clear that sublimation and condensed phase thermolysis compete when NTO is subjected to heating. Thermolysis can be autocatalytic also. There is no consensus regarding the initial decomposition product and hence the decomposition pathway of NTO. Some studies suggested NO_2 as the initial product of decomposition whereas many others reported CO_2 and NO. At least five different mechanisms are suggested on the basis of experimental and theoretical studies. Kinetic parameters measured by various research groups suffer from "kinetic compensation effect". The decomposition pathways and the kinetic parameters are strongly affected by sample characteristics and experimental conditions. No single value for global activation energy is sufficient for the description of the entire thermal decomposition of NTO. NTO salts undergo thermal decomposition in multiple steps. In general, the metal salt hydrates of NTO undergo dehydration initially. This can be in a single or in multiple steps. Then the dehydrated salts undergo exothermic decomposition ultimately yielding metal oxides in most cases or metal carbonates in some cases. Thermal stability of almost all the metal salts is lower than that of NTO, on account of the activation energy for thermolysis and critical temperature for thermal explosion (T_b).

7.1. Introduction

The research in the field of high energetic materials nowadays is directed towards the synthesis of molecules with high energy, high density, high heat resistance and low sensitivity. NTO is just, such an energetic compound with performance properties comparable to that of a high performance explosive compound such as 1,3,5-trinitro-1,3,5-

triazacyclohexane (RDX), while its sensitivity is close to that of 1,3,5-triamino-2,4,6-trinitro benzene (TATB), which is a bench mark insensitive energetic compound.

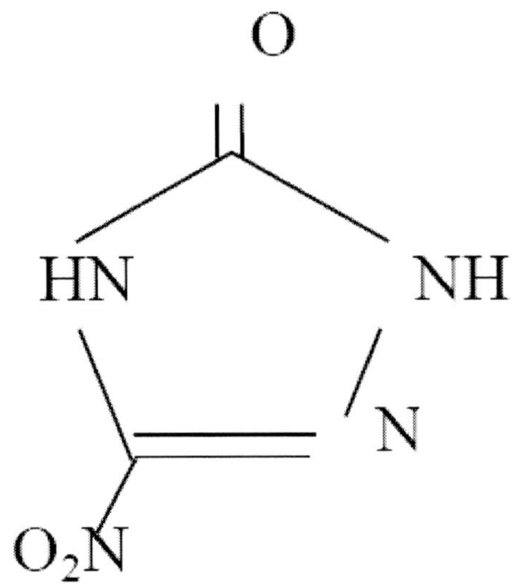

NTO was first reported in the year 1905 from nitration of 1,2,4-triazol-3-one (TO) (the authors have incorrectly attributed NTO to be its hydroxy tautomer) [1]. There was renewed interest in NTO in the mid-late 1960s, but none of these studies appear to be directed to explosives [2-5]. However it was Lee et al., who had first published the study of NTO as an explosive [6]. And subsequently the compound attracted the attention of many researchers worldwide and research on various aspects of NTO has been appeared in a large number of papers in different journals. Singh et al. have recently published a number of reviews on the various aspects of research work done on NTO and its salts [7-10].

A large number of studies directed toward understanding the decomposition pathways of NTO and estimation of kinetic parameters there of have been undertaken [11-25]. Wide ranging techniques such as IR spectroscopy, thermogravimetry (TG), differential thermal analysis (DTA), differential scanning analysis (DSC), TGA coupled with mass spectroscopy (TGA-MS), ARC, NO chemiluminescence, ESR spectroscopy, HPLC and GC analysis of partial decomposition in sealed capillaries and T-jump/FTIR spectroscopy were employed to study the decomposition mechanism of NTO. A number of theoretical studies on the possible decomposition pathways of NTO were undertaken. Controversy still exists regarding the decomposition mechanism of NTO. One school of thought have suggested that the first step in the thermal decomposition of NTO is scission of C-NO_2 bond either by direct thermal activation or catalysed by H atom transfer and loss of HONO [26,27]. The other school of thought has reported that first decomposition products are CO_2 and N_2O [12,17]. Recently, some workers have also reported CO_2 and N_2 as gaseous products while others detected NO_2 and HONO in small amount [23] or not at all [21]. Thus, there is no consensus regarding the decomposition pathway of NTO. Studies on the thermolysis of NTO are discussed in detail in this chapter. Results of both experimental and theoretical studies on thermal decomposition of NTO will be discussed in detail. The various thermolysis pathways put forward by different studies will be critically analysed here.

Owing to the acidic nature of NTO (pK_a 3.67) [6], it forms salts with a large number of metals as well as aromatic and aliphatic amines. The salts of NTO reported in literature can be broadly classified into metal salts and amine salts, on the basis of nature of the compounds with which they are formed. They can also be sub classified as follows:

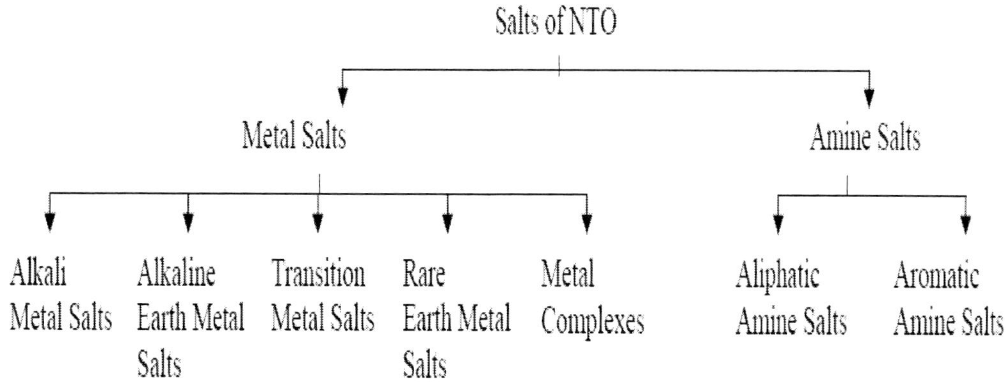

A compilation of all the salts that we have encountered in literature is given in table 1. Since these salts are also insensitive and high energetic in nature, these have also been the subject of intense research interest. A number of papers on preparation, characterisation, thermal and explosive behaviour of the NTO salts have been published. Singh et al. have reviewed various aspects of research on salts of NTO up to the year 2003 [9,10]. Thermolysis of NTO salts is detailed in this chapter so as to get a better understanding of how the pathways of NTO decomposition are influenced by the formation of salts.

7.2. Preparation and Characterisation of NTO

NTO can be prepared by the reaction of semicarbazide hydrochloride (SC) with formic acid to form first triazolone (TO) followed by nitration with 70% nitric acid [4] (Scheme 1).

Crystal packing and molecular dynamics of NTO was thoroughly investigated by Lee et al. [28] and Sorescu et al. [13]. It is known that NTO shows tautomerism and crystalline NTO exists in two phases, α (most stable) and β. It belongs to the triclinic space group P1 and contains eight molecules in the unit cell. However, attempts to determine the coordinates of individual atoms for this structure have not been successful due to "some kind of twinning about the crystal needle axis"[28]. The α-phase has been resolved by X-ray diffraction measurements. The unit cell is monoclinic with four molecules in the unit cell and has the space group $P2_1/c$. The atom labelling of NTO is shown in figure 1 and the geometrical parameters are summarised in table .2.

Table 1. A compilation of salts of NTO reported in literature

Metal Salts					Amine Salts	
Alkali	Alkaline earth	Transition	Rare earth	Complexes	Aliphatic amine	Aromatic amine
$Li(NTO).2H_2O$	$Mg(NTO)_2.8H_2O$	$Mn(NTO)_2.8H_2O$	$Ce(NTO)_3.7H_2O$	$[Sr(NTO)_2(H_2O)_4].2H_2O$	NH_4NTO	RSANTO*
$Na(NTO).H_2O$	$Ca(NTO)_2.4H_2O$	$Fe(NTO)_2$	$Pr(NTO)_3.7H_2O$	$[Yb(NTO)_3(H_2O)_4].6H_2O$	$H_2NCH_2CH_2NH_2(NTO)_2$	
$K(NTO).H_2O$	$Sr(NTO)_2.5H_2O$	$Fe(NTO)_3.2H_2O$	$Nd(NTO)_3.8H_2O$	$[Cd(NTO)_4.Cd(H_2O)_6].4H_2O$	$H_2NNH_2(NTO)_2$	
$Cs(NTO).H_2O$	$Ba(NTO)_2.3H_2O$	$Co(NTO)_2.8H_2O$	$Nd(NTO)_3.7H_2O$	$[Co(H_2O)_6(NTO)_2].2H_2O$	$(NH_2)_2$-C=NH(NTO)	
$Rb(NTO).H_2O$		$Ni(NTO)_2$	$Sm(NTO)_3.7H_2O$	$[Dy(NTO)_2(H_2O)_6].NTO.4H_2O$	$H_2NCNHNH_2(NTO)$ (C=NH)	
		$Cu(NTO)_2$	$Eu(NTO)_3.7H_2O$	$H[Pr(NTO)_4(H_2O)_4].2H_2O$		
		$Cu(NTO)_2.2H_2O$	$Gd(NTO)_3.7H_2O$	$[Y(NTO)_2NO_3(H_2O)_5].2H_2O$		
		$Cu(NTO)_2.4H_2O$	$Tb(NTO)_3.5H_2O$	$[Pr_2(NTO)_4(H_2O)_{10}].2NTO.6H_2O$	$H_2NHNCNHNH_2(NTO)$ (C=NH)	
		$Zn(NTO)_2$	$Dy(NTO)_3.5H_2O$	$[Gd(NTO)_2(H_2O)_6].NO_3.2H_2O$		
		$Ag(NTO).1H_2O$	$Yb(NTO)_3.6H_2O$	$[Ho(NTO)_2(H_2O)_6].NO_3.2H_2O$		
		$Cd(NTO)_2.2H_2O$	$Y(NTO)_3.6H_2O$		$H_2NCNHNH_2(NTO)$ (C=NNH_2)	
		$Hg(NTO)_2.2H_2O$	$La(NTO)_3.7H_2O$			
		$Pb(NTO)_2$				
		$Pb(NTO)_2.H_2O$			$(CH_3)_2NH(NTO)$	
		$Cr(NTO)_3.8H_2O$				

* RSANTO = , where R = H, 2-CH_3, 3-CH_3, 4-CH_3, 2-Cl, 3-Cl, 4-COOH, 2-NO_2, 3-NO_2, 4-NO_2,

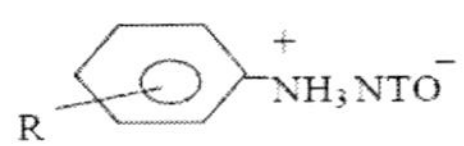

2-OCH_3, 3-OCH_3, 4-OCH_3, 2-OC_2H_5 or 4-OC_2H_5

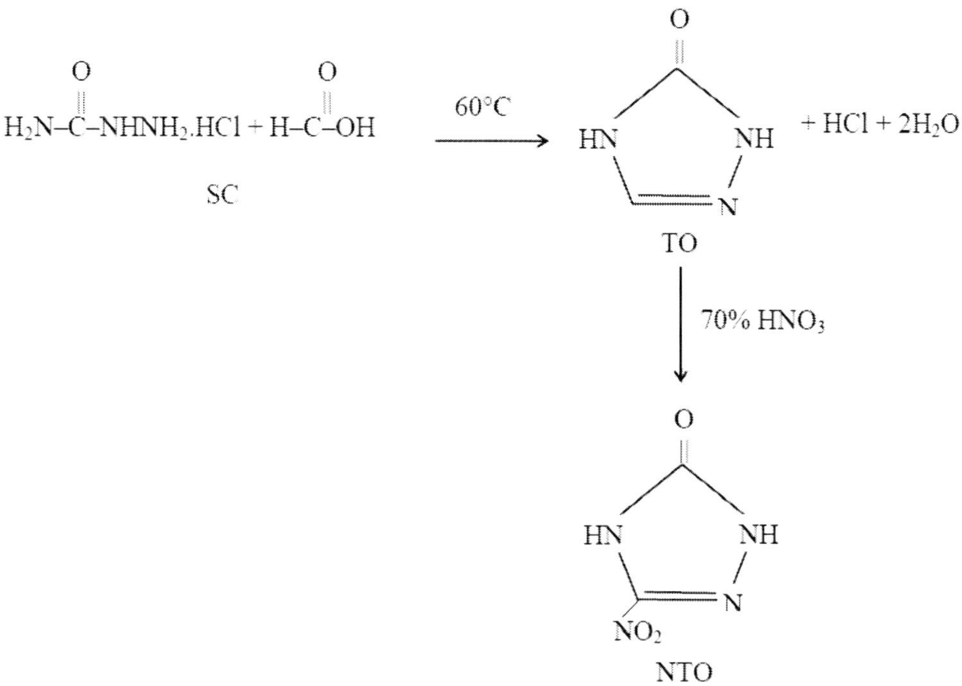

Scheme 1. Preparation of NTO.

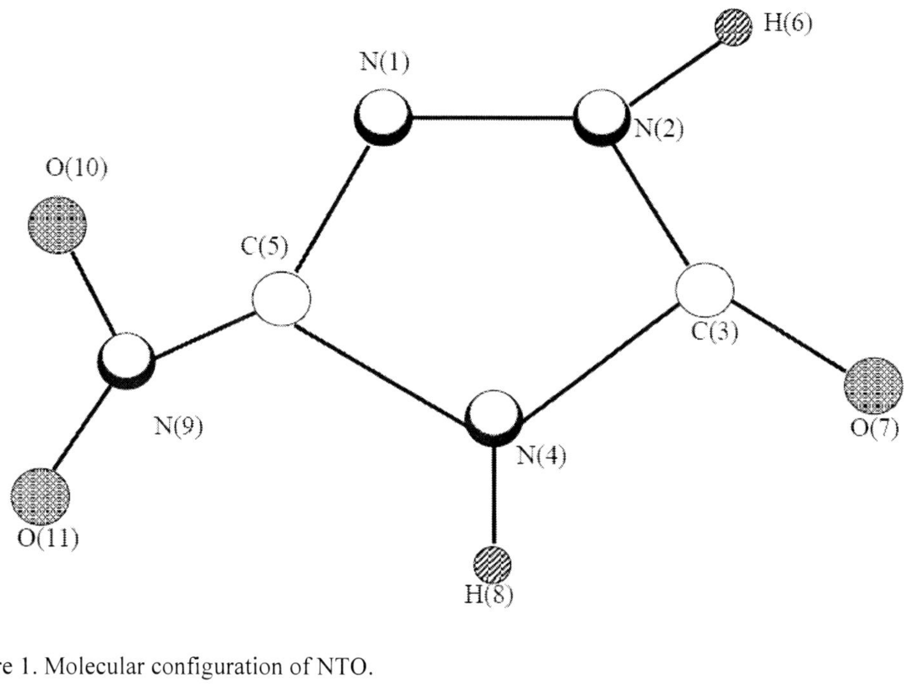

Figure 1. Molecular configuration of NTO.

Table 2. Geometrical parameters for NTO

$r(N_1 - N_2)$	1.369	$\theta(N_1 - N_2 - C_3)$	112.8
$r(N_2 - C_3)$	1.367	$\theta(N_2 - C_3 - N_4)$	103.8
$r(C_3 - N_4)$	1.378	$\theta(C_5 - N_1 - N_2)$	102.4
$r(N_4 - C_5)$	1.349	$\theta(H_6 - N_2 - N_1)$	115.7
$r(C_5 - N_1)$	1.290	$\theta(O_7 - C_3 - N_2)$	126.9
$r(C_3 - O_7)$	1.226	$\theta(O_7 - C_3 - N_4)$	129.3
$r(C_5 - N_9)$	1.447	$\theta(H_6 - N_4 - C_3)$	123.3
$r(N_9 - O_{10})$	1.217	$\theta(N_9 - C_5 - N_4)$	123.2
$r(N_9 - O_{11})$	1.218	$\theta(O_{10} - N_9 - C_5)$	117.4
		$\theta(O_{11} - N_9 - C_5)$	116.2
		$\theta(O_{10} - N_9 - O_{11})$	126.4

Two values are reported for enthalpies of formation of NTO in both gas and solid phases [29, 30]. The values for enthalpies of formation of NTO in gas phase are ΔH_f (g) = -3.2 and -5.7 kcal mol^{-1} and in solid phase are ΔH_f (s) = -30.927 kcal mol^{-1}.

Prabhakaran et al. conducted XRD studies using powder diffraction method and showed that the sample belongs to the tetragonal crystal system with a c/a axial ratio of 0.329 [12]. Crystal structure of NTO has been reported by Sanderson also [31]. Electron density, Laplacian and electrostatic potential distributions of β-NTO have been determined from a low-temperature (100 K) X-ray diffraction experiment [32]. The higher performance characteristics and exceptional insensitivity of NTO is attributed to its special structure [33]. The lone pair of electron on the nitrogen atom joins the conjugation, which results in the enhancement of aromaticity of the ring and thus the thermal stability is increased. The melting point of NTO is increased beyond its decomposition temperature due to the presence of intermolecular hydrogen bonding. A number of theoretical studies on the structure of NTO are available in literature [10]. Further discussion of the theoretical studies on the structure, characterisation and studies on crystallisation of NTO is beyond the scope of this book and hence will not be discussed here. Any interested readers may refer [7-10].

7.3. Preparation of NTO Salts

Preparation of salts of NTO is very easy in most cases, since it only involves simple acid-base reaction. Manchot et al. who were first to prepare NTO in the year 1905, also prepared a potassium ($KNTO.H_2O$) and a mercuric [$Hg(NTO)_2$] salt [1]. Both of the salts were characterised only by elemental analysis. They have also mentioned about the preparation of lithium, barium, sodium and silver salts. Thus the preparation of salts had been reported as long back as the preparation of NTO. However, their potential as high energetic materials has been revealed recently only.

A group of scientists from the Peoples Republic of China reported the preparation and characterisation of a number of metal salts given in table 1 [34-38]. The method of preparation of different categories of NTO salts are discussed briefly.

7.3.1. Alkali Metal Salts

The alkali metal (Li, Na, K) salts were prepared by mixing the hot aqueous solution of NTO with the aqueous solution of corresponding metal hydroxides, and thus salt hydrates were obtained [39,40].

7.3.2. Alkaline Earth Metal Salts

The alkaline earth metal salt hydrates of NTO have been prepared by the addition of corresponding metal carbonate to the hot aqueous solution of NTO with the evolution of CO_2 [41].

7.3.3. Transition Metal Salts

Transition metal salts of NTO (table 1) have been reported to be prepared by the double decomposition of NaNTO/KNTO/LiNTO/NH_4NTO with a suitable salt of the corresponding transition metal in hot aqueous solutions [34, 42].

7.3.4. Rare Earth Metal Salts

Some of the rare earth metal salts of NTO (table 1) have been prepared by treating the corresponding metal hydroxide with hot aqueous solutions of NTO [18]. The preparation by the treatment of aqueous solution of LiNTO with aqueous solutions of corresponding metal nitrates has also been reported [44].

7.3.5. Metal Complexes

$[Dy(NTO)_2(H_2O)_6].NTO.4H_2O$ was prepared by treating the hot aqueous solution of LiNTO and dilute nitric acid solution of dysprosium oxide [45]. $[Yb(NTO_3(H_2O)_4].6H_2O$ and $[Sr(NTO)_2(H_2O)_4].2H_2O$ were prepared by treating hot aqueous solution of NTO with a solution of Yb-oxide/Sr-carbonate [46]. $[Cd(NTO)_4.4Cd(H_2O)].4H_2O$ has been prepared by mixing aqueous solution of NTO and cadmium carbonate in excess [47]. Crystals of $[Co(H_2O)_6](NTO)_2.2H_2O$ were grown from aqueous solution containing $Co(NO_3)_2$, NTO and LiOH [48]. The metal complex of Pr, $H[Pr(NTO)_4(H_2O)_4].2H_2O$ was prepared by mixing the aqueous solution of LiNTO and the dilute nitric acid solution of Pr_6O_{11} [49]. Another complex of the same metal but with different composition $[Pr_2(NTO)_4(H_2O)_{10}].2NTO.6H_2O$ from the same starting materials (LiNTO and dilute nitric acid solution of Pr_6O_{11}) was formed by slow crystallization at low temperature [50]. Complexes of actinides such as Gd and Ho of the composition, $[Gd(NTO)_2(H_2O)_6].NO_3.2H_2O$ and $[Ho(NTO)_2(H_2O)_6].NO_3.2H_2O$ were also formed by slow crystallization from a mixture of aqueous solution of LiNTO and dilute nitric acid solutions of the respective metals [50]. $[Y(NTO)_2NO_3(H_2O)_5].2H_2O$ was made from the mixture of aqueous solution of LiNTO and dilute nitric acid solution of yttrium oxide [51].

7.3.6. Aliphatic and Aromatic Amine Salts

The amine salts of NTO have got a pretty long history as Chipen et al. reported the preparation of some amine salts along with the preparation of NTO and its metal salts [3]. But detailed study of preparation, characterisation and application started with the joint efforts of US department of the army and the US department of energy. The preparation of the amine salts of NTO were accomplished by mixing the hot aqueous solution of NTO with the solution of corresponding base. Lee et al. have prepared seven amino salts of NTO [Hydrazine (HNTO), ammonia (ANTO), ethylenediamine(ENTO), guanidine(GuNTO), aminoguanidine(AGuNTO), diaminoguanidine (DAGuNTO), triaminoguanidine (TAGuNTO)], and characterised them using ^{13}C-NMR spectroscopy [52,53]. Singh et al. have also prepared fifteen ring substituted arylammonium salts of NTO [54].

7.4. Characterisation and Structure of the Salts of NTO

Almost all of the reported salts of NTO have been well characterised by ^{13}C-NMR, IR, microanalysis and XRD [39-54]. Thermal behaviour and energetic parameters have also been assessed in most cases. Earlier workers have not studied the structural properties in detail and the molecular formulae have been identified simply from elemental analysis. There are two ionisable hydrogen sites in NTO molecule at positions N(2) and N(4). Redman et al. argued that both the positions are easily ionisable and thus NTO form salts of the type $M^{n+}[NTO\text{-}H]^-$ or $M^{n+}[NTO\text{-}2H]_{n/2}^{2-}$ [eg. $Hg(NTO\text{-}H)_2$, Hg(NTO-2H)] [39]. Both acidic and normal salts of NTO have been reported with the same metal. But their observations have been contradicted later on, by the Chinese research workers on the basis of their spectral and XRD studies on salts of NTO [41].

7.4.1. ^{13}C-NMR Studies on Salts of NTO

^{13}C-NMR is a powerful tool to characterise NTO as well as its salts. The values of chemical shift (δ, ppm) from ^{13}C-NMR of NTO and its salts has been summarised in table 3. ^{13}C-NMR was used to determine the site from which the proton leaves when NTO is converted into some of its salts [33]. The spectrum of NTO gives two values for chemical shifts, approximately at 148 ppm and 154 ppm respectively, which are reported to be corresponding $>C\text{-}NO_2$ and $>C{=}O$ [6]. Comparing the ^{13}C-NMR spectrum of NTO with some of its salts, it has been observed that the chemical shifts of both the carbon atoms in the NTO salts are to the downfield and that the amount of shifting is almost the same [33]. This indicates that the attraction is equal to both the carbon atoms and for this to occur the proton might leave from position N(4) which is in between the two carbon atoms. Moreover, our compilation in table 3 shows that this observation could be extended to almost all other salts of NTO with very few exceptions (rare earth metal salts). It is also notable that some metal salts of NTO like CuNTO does not show any peak in ^{13}C-NMR spectrum due to paramagnetism [33].

Table 3. ^{13}C-NMR data of NTO and its salts

S. No.	Compound	Solvent	Chemical Shift , δ (ppm)		Ref
			>C-NO_2	>C=O	
1.	NTO	DMSO-d_6	148.00	154.40	[6]
		$CDCl_3$	148.13	154.70	[33]
2.	ANTO	DMSO-d_6	159.40	164.60	[53]
3.	ENTO	DMSO-d_6	158.10	164.40	[53]
4.	HNTO	DMSO-d_6	159.40	164.86	[53]
5.	GuNTO	DMSO-d_6	159.12	164.86	[53]
6.	AGuNTO	DMSO-d_6	159.33	164.86	[53]
9.	DAGuNTO	DMSO-d_6	159.30	164.96	[53]
10.	TAGuNTO	DMSO-d_6	160.00	165.00	[53]
11.	KNTO.H_2O	$CDCl_3$	159.97	165.17	[33]
12.	$Mg(NTO)_2.8H_2O$	DMSO-d_6	159.77	165.04	[41]
13.	$Ca(NTO)_2.4H_2O$	DMSO-d_6	159.97	165.23	[41]
14.	$Sr(NTO)_2.5H_2O$	DMSO-d_6	159.25	165.04	[41]
15.	$Ba(NTO)_2.3H_2O$	DMSO-d_6	159.83	165.04	[41]
16.	PbNTO	DMSO-d_6	159.19	164.52	[33]
17	$Pr(NTO)_3.9H_2O$	DMSO-d_6	146.51	172.74	[43]
18	$Nd(NTO)_3.9H_2O$	DMSO-d_6	145.46	166.36	[43]
19.	$Sm(NTO)_3.8H_2O$	DMSO-d_6	150.73	156.00	[43]

7.4.2. X-Ray Diffraction Studies on Salts of NTO

Single crystals were grown and structure has been determined using XRD in many cases. The crystal structures of ENTO [55] and DAGuNTO [56] were resolved by X-ray crystallography. Crystals of ANTO were grown by slow evaporation from aqueous solution and the structure is orthorhombic with space group P_{222} [14]. The geometrical structure of ANTO has also been investigated using semi empirical molecular orbital (MO) calculations, which revealed that there are four distinguished H-bondings in the ANTO molecular or ionic system [57]. This structural determination has also indicated that the N(4) proton is more acidic than the one at N(2) and thus deprotonation is easier from N(4), and this result is in agreement with the results reported by Ritchie [29]. The deprotonation of N(4) is also confirmed by X-ray analysis of some crystalline metal salts [38,58]. The crystal structures of all the metal complexes [45-48, 50,51] given in table 1 is monoclinic except that of $H[Pr(NTO)_4(H_2O)_4]2H_2O$ [24] and $[Ho(NTO)_2(H_2O)_6].NO_3.2H_2O$ [50] which are triclinic.

7.4.3. IR Spectral Studies on Salts of NTO

In the case of alkaline earth metal salts of NTO, IR spectra show that, when NTO is converted into its salts, the absorption frequencies for >C=O and –NO_2 groups as well as for the frequencies of C-N, C=N, C=O and N-H are all shifted to lower wave numbers. These results indicates that the negative charge on the NTO ring decreases when NTO co-ordinates with a metal cation. The far IR spectrometric results confirm that the metal cation co-ordinates with both nitrogen and oxygen atoms [41].

7.5. EXPLOSIVE BEHAVIOUR OF NTO AND ITS POTENTIAL APPLICATIONS

The major attraction of NTO as an energetic material is its insensitivity and good explosive performance. Although many of its performance parameters are available in literature, most of the studies on the applications of this compound are hidden in classified literatures. We have collected the reports available in open literature and an overview is presented here.

There are many detailed studies available on various performance parameters and sensitivity characterization of NTO [1,31,59-66]. The sensitivity data from various references are summarized in table 4. It can be seen from the table that the same properties reported by different research groups differ considerably, which may be attributed to the difference in experimental methods. The data for RDX are also presented alongside to that of NTO for comparison. Some experimental results for explosive performance of NTO have been summarized in table 5. Recently Agrawal et al. have conducted a high-speed photographic study of the impact initiation mechanisms of hexanitro-hexaaza-isowurtzitane (CL-20) and NTO by drop weight impact [66]. NTO was found to be less sensitive than β-HMX and CL-20. But it was also found to be sensitized by both hard high melting grits and brittle polymers.

Table 4. Sensitivity of NTO and RDX to various hazardous stimuli

Test	NTO	RDX	Ref
Impact sensitivity			
Rotter Figure of Insensitivity (mean gas evolved, ml)	90(3), 80(5)	80(17)	[67]
$h_{50\%}$ (cm)	291	22	[1]
Energy (J)	22	4.5	[52]
Thermal Sensitivity			
Vacuum Thermal Stability			
ml /g / 120 ^{0}C / 40h	0.0	0.2 – 0.6	[67]
ml /g / 120 ^{0}C / 40h	0.3	0.12 - 0.9	[1]
Ignition Temperature			
Temperature of Ignition (^{0}C)	258	216 ± 3	[67]
Henkin critical temperature (^{0}C)	237	219.6	[1]
Temperature of Ignition (^{0}C)	280	220	[52]
Temperature of exothermic peak			
DSC (^{0}C)	273	212	[67]
DTA (^{0}C)	> 236	210	[1]
Spark Sensitivity			
Energy (J)	> 4.5 (no ignition)	4.5 (ignition)	[67]
Energy (10 mil, J)	3.40	0.55	[1]

Table 5. Performance parameters reported for NTO

Charge Density (g/cm^3)	% TMD	Charge diameter (cm)	VOD (m/s)	P_{CJ} (GPa)	Ref
1.781	92.2	4.13	--	27.8	[1]
1.853	96.0	4.13	--	26.0	[1]
1.782	92.3	2.54	--	24.0	[1]
1.855	96.1	2.54	--	Failed	[1]
1.759	91.1	1.27	--	25.0	[1]
1.824	94.5	1.27	--	failed	[1]
--	89.61	1.27	6040	--	[67]
--	88.65	1.59	7440	--	[67]
1.69	--	--	7400	--	[60]
1.71	--	--	7770	--	[60]
1.91	--	--	8590	--	[61]
1.91	--	--	8200	33.8	[62]
1.91	--	--	8590	--	[63]
1.81	--	1.45	No go	--	[65]
1.80	--	1.60	7650	--	[65]
1.81	--	2.00	7800	--	[65]
1.80	--	2.50	7820	--	[65]
1.80	--	3.00	7860	--	[65]
1.78	--	2.36	7800	24.6	[65]

Spear et al. have formulated plastic bonded explosives with four different polymers (three ethylene vinyl acetate-EVA copolymers and a polyphosphazine) in a NTO: polymer ratio of 95:05 [67]. The molding granules of NTO formed with EVA polymers were found to be highly insensitive to impact. In contrast, NTO/polyphosphazine was slightly desensitized relative to NTO, but propagation ability was slightly decreased. Huang and Wu described a method for estimating the autoignition temperature of NTO and its salts from DSC data [68]. Volk and Bathelt have calculated the energy output of NTO in gun propellants using ICT-Thermodynamic code [69]. Fouche et al. have reported synthesis, recrystallisation, chemical and physical characterization of NTO [70]. They have studied various RDX-TNT, NTO-TNT and RDX-NTO-TNT formulations. The studies proved that, incorporation of NTO and/or fine RDX in TNT containing munitions, showed significant improvements in sensitivity, mechanical properties and structural integrity of explosive columns. They have concluded that 40:60 NTO:TNT and 25:25:50 RDX:NTO:TNT can be considered safe and reliable charge for various caliber artillery shells. Kong et al. have studied the temperature dependence on composition and constructed phase diagram for 85:15 NH_4NO_3-KNO_3/ethylenediamine dinitrate/NTO [71].

Sunrall has developed a melt-castable general-purpose insensitive high explosive formulation of NTO with a thermoplastic binder (TTB-531) [72,73]. The formulation also contained aluminium powder, ammonium perchlorate (AP) and RDX. The sensitivity and performance tests on this formulation demonstrated that it is possible to formulate a high-performance melt-castable explosive that does not detonate during either fast-cook off or slow-cook off situations. Ellis and Benzuidenhout have determined detonation energy, ballistic and Gurney energies of various NTO based formulations [74]. Fouche et al. have developed NTO based pressed PBX-formulations [75]. They have found that HNS/Kel F could be used as a suitable booster formulation for the reliable initiation of a pressable main charge filling consisting of NTO/RDX/EVA. Moreover, NTO/RDX/EVA is suitable for initiation of NTO/TNT melt castable main charge filling. Thus they have constructed a reliable explosive train. Mukundan et al. have recently explored NTO as a substitute for RDX in some munitions [76].

Williams et al. have shown that NTO on thermolysis leaves a melon type cyclic azine polymeric residue, which is thermally stable up to 700 ^{0}C [21]. Therefore they have suggested that NTO could be used as a potential burning rate additive in solid propellants to suppress the burning rate and enhance combustion stability. Dalin et al. have shown that pyrolysis characteristics of NTO and its salts could be used as a criterion to evaluate their catalytic activity, when they are used as ballistic modifiers [77]. However a recent study by Singh et al. showed that NTO affects the processing parameters by building up viscosity when it is used as a ballistic additive in hydroxyl terminated polybutadiene (HTPB)-AP based composite solid propellants [78]. However it has also been shown that NTO slightly increases the burning rate of the said propellant at 2 % by mass concentration. Another potential application suggested and tested for NTO is in propellant compositions for automobile restraint air bag systems [79-81]. NTO and its salts are used in these systems in combination with other energetic additives.

7.6. Thermolysis of NTO

Of the various parameters, thermal stability is the most important safety aspect of an energetic material, since any hazardous stimuli, eventually triggers off a thermal event, which is the main cause of initiation of any explosive [82]. The early thermal decomposition reactions are very much important from the viewpoint of understanding the mechanism of explosion. During thermolysis of NTO sublimation process competes with decomposition in the condensed phase [18, 25]. Since the decomposition reaction steps depend upon temperature, pressure and possible phases, studies carried out under different experimental conditions have led to strong disagreement about the initial pathways. As briefly mentioned in the introduction of this chapter, a wide variety of instrumental techniques were used for studying the NTO decomposition. In addition to thermoanalytical techniques, photochemical and laser induced decomposition studies were undertaken. At least five different decomposition routes are suggested for NTO. We will discuss these decomposition pathways here along with the reported kinetic parameters. The instrumental technique used for arriving at a particular conclusion will also be mentioned.

7.6.1. Experimental Methods and Proposed Initial Steps

Homolysis of one or more bonds to ring substituents: Many of the studies suggested that the first step in thermal decomposition of NTO is the scission of C-NO_2 bond, either by direct thermal activation or catalyzed by H atom transfer. Beard and Sharma have subjected NTO to X-ray and UV radiation damage and they have used X-ray photoelectron spectroscopy (XPS) to detect the damage induced [83]. The photo-damaged residues suggested the loss of nitro functionality and/or its transition to a more reduced chemical state. A comparison of the sensitivity of NTO, RDX, HMX and TATB to X-ray and UV radiation was also made. The order of sensitivity was as follows: RDX > HMX > NTO >TATB. It was also concluded that NTO is unique among commonly used energetic materials in that it may not be sensitized by molecular degradation. Beard and Sharma have also studied the early reaction chemistry of NTO, induced by drop weight, impact, shock, heat and radiation, using XPS and chemical ionization mass spectrometry (CIMS) [84]. XPS spectra indicated the loss of NO_2 concentration in response to damage in all cases.

Ostmark et al. have studied the decomposition of NTO using MS [both electron impact (EI) and CI], laser induced-MS (LIMS), chemiluminescence and DSC. [11,23,24]. The mass spectrum consisted of three characteristics parts: the molecular ion (m/z 130), the groups at m/z 83, 84 and 85 (the azole ring) and the azole ring fragment groups at m/z 41, 42, 43 and 44. The fact that the molecular ion (m/z 130) was found in relatively large amount in the electron impact (EI) spectrum (70 eV) indicates that the NTO molecule is very stable compared with HMX, RDX and TNT. The chemical ionisation (CI) spectrum of NTO showed two peaks at m/z 131 and m/z 261 (2M + 1); 261 peak is probably a cluster ion (adduct). The mass spectrum for TO (EI 70 eV) consisted of only two peaks, at m/z 42 and 85. The proposed mass spectral fragmentation of NTO is shown in Scheme 2. They have concluded that the probable thermal decomposition mechanism is the elimination of NO_2 followed by a breakdown of azole ring.

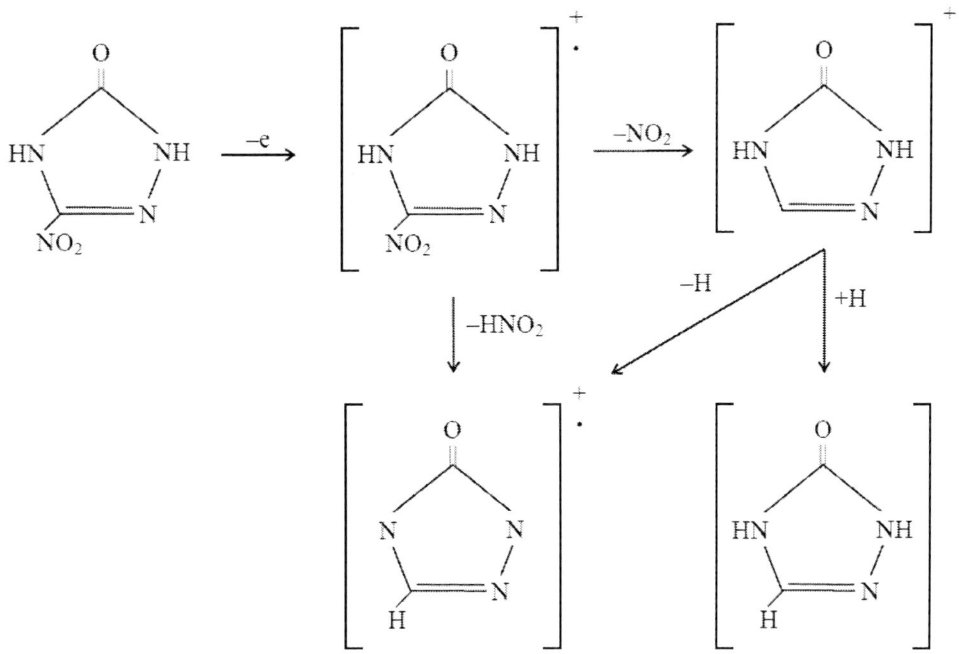

Scheme 2. Proposed fragmentation pathways corresponding to the observed mass spectral signals.

Prabhakaran et al. [12] have studied the mechanism of thermolysis of NTO using TG, DTA, IR, DSC, XRD and hot stage microscopy. They have analysed the evolved gases during NTO decomposition by IR spectroscopy. At around 244 ^{0}C, carbon dioxide and nitrous oxide were detected as early gases. The bands due to CO_2, NO_2 and N_2O were prominent in the spectra. Formation of $^\bullet NO_2$ radical by C-NO_2 cleavage, with rupture of the adjacent C-N bond appears to be the probable mechanism in the thermal decomposition of NTO. Their results show that cleavage of C-NO_2 and the oxidative attack of the NO_2 group on the ring fragment accounts for the formation of gaseous products such as CO_2, NO and N_2O and the following mechanism (Scheme 3) was proposed:

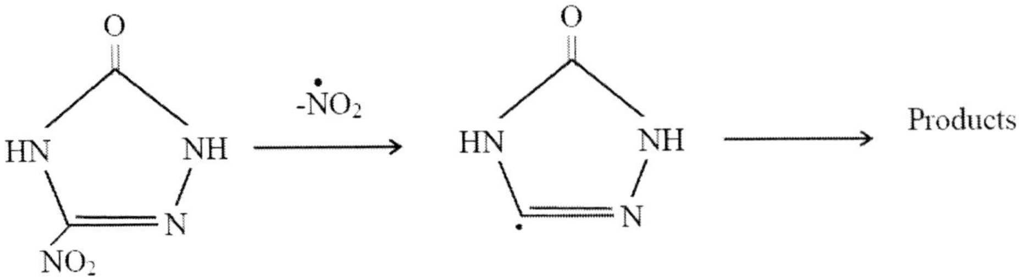

Scheme 3. NTO thermolysis pathway suggested by IR spectroscopic detection of decomposition products.

It was Williams et al. who first refuted the hypothesis that C-NO_2 homolysis is the initial step in thermolysis of NTO [21]. They investigated the product gases of NTO decomposition under fast heating rates, using IR spectroscopy, as a function of pressure and found neither NO_2 nor HONO initially. Oxley et al. studied the decomposition of NTO, over a wide range of temperature (220-280 ^{0}C) [19]. They have proposed a high temperature mechanism, which is initiated by loss of HONO or NO_2. The autocatalytic nature of the NTO decomposition is attributed due to the generation of $^{\bullet}NO_2$ free radical. Decomposition in solution or in open thermolysis tubes is strictly of the first order. Analysis indicated the presence of $C_2H_2N_3O$, a formula consistent with "polymerised TO" (1,2,4-triazole-5-one). TO is not a direct intermediate in the formation of this residue. Independent decomposition of TO resulted in a residue of somewhat different elemental composition [$C_{4.5}H_5N_{7.7}O$] as reported in Scheme 4. They have also reported both primary and secondary deuterium kinetic isotope effects (DKIE) and concluded that hydrogen transfer is involved as a rate-limiting process at lower temperatures. A subsequent study by Oxley et al. suggested two distinct mechanisms [85]. One pathway proceeds by means of CO_2 and N_2 evolution, a route that would account for the observed DKIE [15,19]. They have also speculated that, nitro group migration from the ring carbon (C_5) to an adjacent ring nitrogen (N_4) may occur. Oxley et al. tracked NTO decomposition products, using ^{15}N labels and this study also asserted the mechanism-involving homolysis of nitro group from the NTO ring [16].

Rupture of triazole ring: Beardell et al. have studied the thermolysis of NTO using pulsed IR laser pyrolysis of thin films and Fourier – transform IR spectroscopy, for product analysis [86]. Unlike previous investigators, they observed appreciable quantities of CO_2, and was suggested to be formed by intramolecular oxidation by the nitro group at the keto carbon atom. Moreover, they detected neither NO_2 nor HONO as primary decomposition species. They suggested a ring opening mechanism.

Exchange of the nitro substituent with vicinal hydrogen substituent: Oxley et al. have proposed that the exchange of the nitro group with vicinal hydrogen substituent, yielding higher energy intermediate as an initial step [85]. This exchange is followed by a unimolecular breakdown of triazole ring, yielding CO_2 and other products.

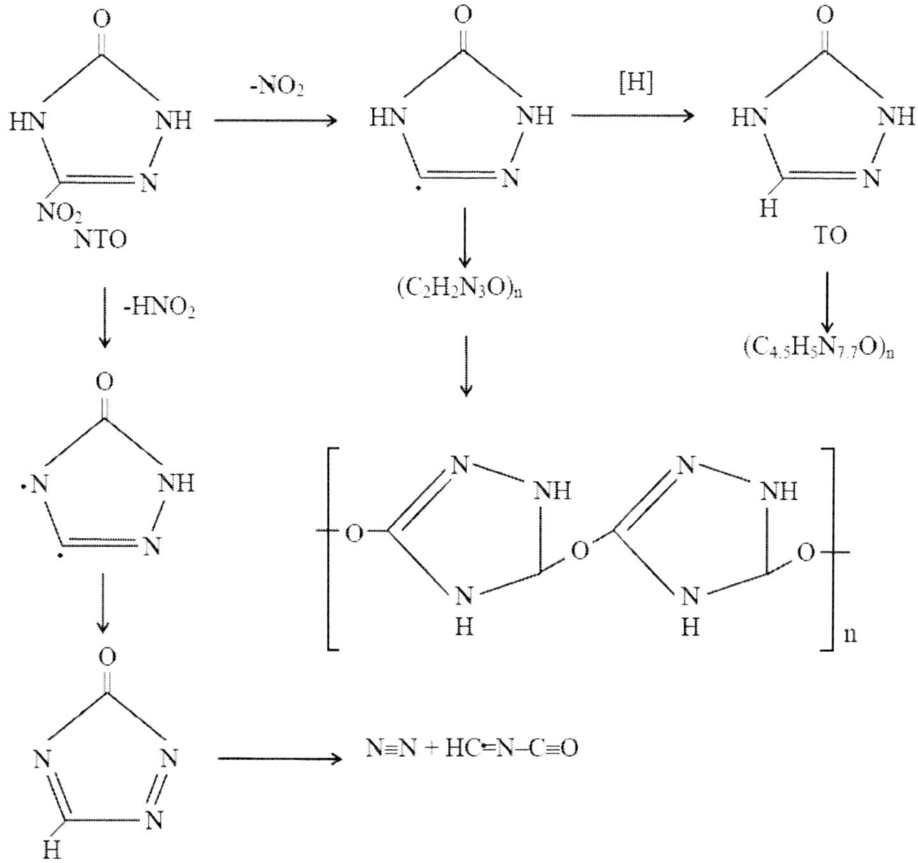

Scheme 4. Autocatalytic thermolysis pathway suggested for NTO due to the formation of free radicals.

Nitro-Nitrite rearrangement: McMillen et al. obtained real time photoionisation of NTO, using surface analysis by laser ionization (SALI) apparatus [22]. Their result suggested several decomposition pathways, all of which feature a nitro-nitrite rearrangement followed by NO loss. Using vacuum UV single-photon ionization, the shear induced NTO spectra were obtained with a spring-driven shearing device installed in a SALI chamber directly beneath the mass spectrometer sampling region. The shear induced spectra were dominated by a peak at m/z 99, which is not seen in previous thermal studies. This peak was assigned to the closed-shell triazole-diketone produced by a nitro-nitrite rearrangement, followed by NO loss and then by rapid bimolecular H-atom removal. The stability of the cyclic diketone intermediate thus generated could help to explain the shock insensitivity of NTO. Laser-desorption spectra were also obtained both on fresh NTO samples and samples that have been recovered from marginally sub-critical drop weight impact tests. The sequence of thermally generated fragments were mainly m/z = 16, 30, 45, 46 and 59. This series suggests several decomposition pathways, dominated by the same nitro-nitrite rearrangement and NO loss as the shear induced decomposition. However, under the thermal or laser-desorption conditions, subsequent bimolecular H-atom removal to produce the closed-shell diketone is evidently slower than unimolecular ring opening adjacent to the carbonyl group. It was shown how this sequence satisfactorily explains:

i. the initial formation of CO_2 that has been previously reported.
ii. the result of nitrogen double-labelling experiments.
iii. the fact that neither NO_2 nor HONO have been seen as substantial initial products of NTO decomposition.

Bimolecular routes: Many bimolecular decomposition routes have also been suggested for the thermolysis of NTO. One of the bimolecular mechanisms was reported by Menapace et al. from their room temperature EPR photochemical decomposition experiments on NTO ($NTO-d_2$)/acetone (acetone-d_6) [15]. On the basis of the obtained EPR spectra three radical species (I, II and III) can be identified as shown in figure 2. The observation of identical EPR spectra in the NTO/acetone, NTO-d_2/acetone and NTO/acetone-d_6, decomposition corresponding to radicals I-II, implies that two different mechanistic reaction pathways are operative leading to the formation of radicals I-III. One reaction pathway involves an intermolecular reaction between two NTO molecules, and the other involves a solvolysis reaction pathway between NTO and acetone.

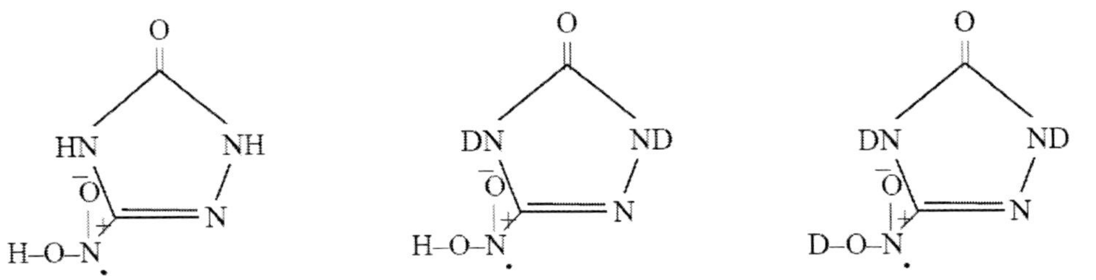

I
hydroxy 4–oxo–2, 3, 5–triazolyl nitroxide

II
hydroxy 4–oxo–2, 3, 5–triazolyl –d_2 nitroxide

III
hydroxy–d– 4–oxo–2, 3, 5–triazolyl d_2–nitroxide

Figure 2. Identified radicals from EPR photochemical decomposition studies of ($NTO-d_2$)/acetone (acetone-d_6).

The low temperature (370 K) EPR thermochemical decomposition performed on the NTO ($NTO-d_2$)/TNT ($TNT-d_5$) mixtures yield EPR spectra of radical intermediates which can be assigned as IV, V, VI and VII species as shown in figure 3.

The overall mechanistic pathways can be summarised as given in Scheme 5. The pathways appear to involve reactions by which the NO_2 group on NTO abstracts hydrogen atoms from other NTO species and/or from acetone.

Botcher et al. have confirmed the previous identification of CO_2 as the initial gas-phase product from their NTO thin film studies [87]. The kinetic data in this study suggested an initial bimolecular step. Thus oxygenation of carbonyl group to CO_2 was found to be occurring by reaction with NO_2 functional group on a neighboring molecule.

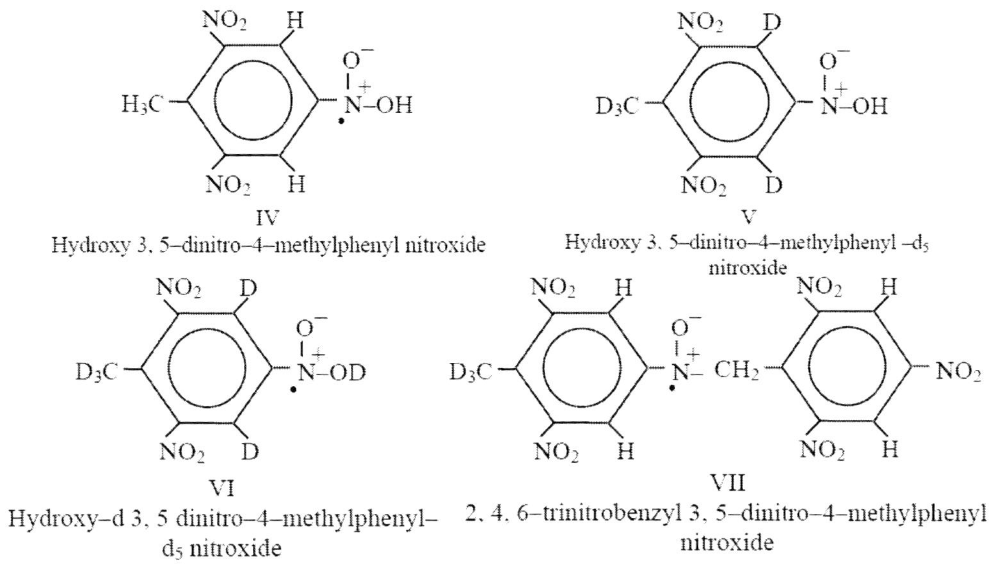

Figure 3. Identified radicals from EPR thermoemical decomposition studies of NTO (NTO-d_2)/TNT (TNT-d_5) mixtures.

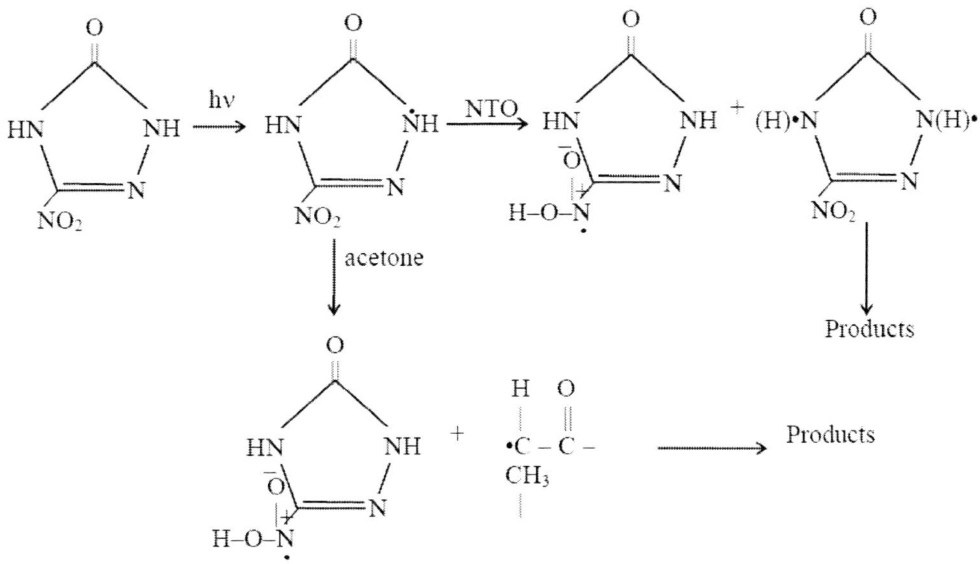

Scheme 5. The reaction pathways suggested for the photothermal decomposition of NTO from EPR studies.

Garland et al. studied the decomposition of NTO using photo-thermally heating solid NTO samples by irradiating with pulsed laser light and the gaseous products were detected using single-photon ionization on a time of flight MS [17]. The relative yields of products with m/z = 114, 101, 85 and 83 were measured as a function of temperature. The product yields were normalized by the velocity-weighted, integrated ion signal from the parent NTO molecule. The early reaction pathways were characterised by analysis of temperature dependence of the relative products by net loss of an oxygen atom from NTO. The results were consistent with earlier observation of evolution of CO_2 as an early product. They suggested the initial reaction to be bimolecular with a net loss of O atom to yield nitroso-TO. However, they also noted that fragments produced by loss of NO and by scission of the C-NO_2 bond are early, but not the initial products.

7.7. KINETICS OF THERMOLYSIS OF NTO

Measurement of the kinetics and associated Arrhenius parameters of thermal decomposition is one of the important characterization of any energetic compound. This is important for prediction of safety and performance parameters. Kinetics of thermolysis of NTO has been measured by many research workers and some of these studies are discussed here.

Thermal decomposition of NTO was studied using DSC, TGA-MS and ARC by Rothgery et al. and the major gaseous products were identified [26]. They have observed that the stability of NTO is sensitive to ageing of the sample. Kinetic analysis suggested an auto-catalytic mechanism. The decomposition mechanism changed abruptly at 212 ^{0}C where an ignition occurred. Under identical conditions, the onset temperature for an NTO thermal runway was approximately 28 ^{0}C higher than that found for RDX and its ignition temperature was approximately 12 ^{0}C higher [88]. Prabhakaran et al. have also studied the kinetics of the isothermal decomposition in the temperature range 229-246 ^{0}C under static air atmosphere using model-fitting method [12]. Out of the eighteen equations tried, the best linearity, with a correlation coefficient of 0.9937 was obtained with Avrami-Erofeev equation for the range of 0-83 % by mass decomposition that gave activation energy of 44.8 kcal mol^{-1}. The thermal stability of NTO was studied by a chemiluminescence (CL) method [11,23]. Equipment using the chemiluminescence reaction between nitric oxide and ozone was used primarily for studying NO development from explosives at different temperatures. Assuming zero order Arrhenius kinetics, one can estimate the frequency factor and activation energy. From CL measurements the activation energy for NTO was calculated as 140 kJ mol^{-1} and the frequency factor as 5×10^6 s^{-1} (temperature range = 100-140 ^{0}C). Other techniques such as IR, HPLC and TG-DTA were also used to measure global kinetic parameters for the thermal decomposition of NTO. Brill et al. have compiled all the global kinetic parameters for the thermal decomposition of NTO that is reported by various workers using different techniques and is given table 6 [20]. They observed the existence of "kinetic compensation effect", i.e. a linear relationship between reported values of activation energy (E_a) and frequency factor (lnA) [$\ln A = a E_a + b$, a and b are compensation factors having values 1.03 mol/(kcal s) and -13 ln k/s^{-1} respectively].

Table 6. Kinetic parameters reported for NTO decomposition

Ea / (kcal mol^{-1})	lnA / s^{-1}	Temperature range / ^0C	Instrument used	Ref
25.8	29.2	134-163	isothermal TGA	[18]
28.6	31.1	134-175	nonisothermal TGA	[18]
31.1	30.1	180-230	nonisothermal TGA	[18]
40.7	26.9	100-137	NO chemiluminescence	[23]
44.8	38.3	229-246	isothermal TGA	[12]
49.3	47.6	195-210	IR	[12]
78.6	67.0	220-280	HPLC, solution	
87.1	74.8	289-299	T-jump	[18]
87.5	77.8	225-245	HPLC	
120.4	112.1	539-553	DSC	[33]

The Brill group has further analysed the kinetics of NTO and observed that the reported values of Ea fall into two ranges: 40.7-52 and 78-120 kcal/mol [18]. In order to explain this they undertook further decomposition studies under slow as well as fast heating experiments on NTO and obtained the kinetic parameters under different conditions [18]. The T-jump/FTIR experiment was employed to determine the kinetic constants for decomposition of flash-heated NTO. They found that two dominant processes occur when NTO is heated at elevated temperatures: sublimation and decomposition. By optimizing the conditions that are preferential to sublimation (i.e., low pressure, moderate temperature), the rate constants especially for sublimation were determined for the first time. The formation of sublimed NTO without detectable solid or gaseous decomposition products were analysed by IR. The kinetic constants for decomposition that yield E_a = 45-52 kcal mol^{-1} almost entirely represent sublimation of NTO in the lower temperature range and a mixture of decomposition and sublimation at higher temperatures. In the unconfined state, the kinetic constants may be a mixture of the rates of sublimation and decomposition, the balance of which shifts with changes in two dominating processes may be separated and independently investigated. If the thermal decomposition rate of NTO is desired, then the kinetics which yields activation energies in the vicinity of 78-87 kcal mol^{-1} rather than 45-52 kcal mol^{-1} should be used.

G.T. Long et al. have argued that no single value of E_a can adequately describe the entire decomposition process of NTO [25]. More generally speaking, activation energies for solid state reactions change during the course of reaction where multiple competing physical and chemical processes can occur simultaneously. Hence it is recently being argued that a more accurate means of characterisation of solid-state processes would employ a method that yields activation energies as a function of the extent of reaction specified as E_α. Thus, Long et al. have used model-free isoconversional analysis for obtaining the dependencies of E_α on extent of reaction, α. They used TG and DSC techniques under conditions selectively promoting the competing reaction channels. Sublimation occured more readily in an open pan and a global activation energy of 130-140 kJ mol^{-1} was obtained for sublimation of NTO. Note that this value is close to the values obtained by the Brill group. Non-isothermal TGA and DSC thermal curves run on pierced pan provided evidence for competitive sublimation and condensed-phase decomposition of NTO. Confining NTO samples in a closed pan resulted in condensed-phase decomposition that leads to the formation of gaseous products and show autocatalytic behaviours during the latter stages. Isoconversional analysis of DSC thermal curves of closed pan experiments yielded activation energies for exothermic thermolysis that increases from $E_\alpha = 273$ kJ mol-1 for $\alpha = 0.01$ to a plateau of 333 kJ mol-1 for $0.17 \leq \alpha \leq 0.35$

prior to decreasing to 184 kJ mol-1 for α = 0.99. The decrease in E_α with α during the latter stage corresponds to autocatalytic behaviour. Thus, this study has quantified activation energies for all three major processes, i.e. sublimation, decomposition and autocatalysis for NTO.

7.8. THEORETICAL STUDIES ON THERMAL DECOMPOSITION OF NTO

Although considerably large number of studies have been reported on the thermal decomposition of NTO, there is no assertive study regarding the mechanism of the process. Theoretical studies based on various quantum mechanics (QM) calculations may be of help, in further understanding of the mechanism. Ostmark et al. have done some empirical, QM calculations, in order to explain the high stability of NTO and verify the initial step in the decomposition path [11]. The bond scission energy, ΔE* for C-NO_2 and the two N-H bonds were found to be 255, 328 and 345 kJ mol^{-1}, respectively. Thus these calculations verified that the week bond in NTO is the nitro C-N bond and not the N-H bond.

Harris et al. have studied bond dissociation of NTO also along with their theoretical studies on tautomerism and ionization [89]. They have estimated N-H and C-NO_2 bond dissociation energies for the planar keto tauomer as 389 and 293 kJ mol^{-1}, respectively. Thus they proposed that C-NO_2 bond cleavage is the likely initial decomposition step at high temperatures, whereas hydrogen atom transfer may play a key role in the condensed phase. Moreover, since several tautomers of NTO are energetically accessible, these may prove to be significant in the condensed phase, where they can be formed by a base catalyzed mechanism.

Meredith et al. have used *ab initio* QM methods to study the energetics associated with several proposed initiation routes of NTO [90]. A comparison of the energies computed for NTO and other molecules that may be formed during its decomposition provided considerable insight into the feasibility of the proposed unimolecular decomposition mechanisms for NTO. The results also indicated that homolysis of C-NO_2 bond (energy = 280.3 kJ mol^{-1}), is the unimolecular pathway, which require lowest initial energy. For the two-step mechanism involving loss of HONO, 313.8 kJ mol^{-1} of energy is required. Thus on the basis of the energetics of the various schemes, C-NO_2 bond homolysis was found to be the most plausible initial step for unimolecular decomposition of NTO.

Wang et al. have considered 39 decomposition paths among 18 intermediates and 14 transition states [91]. They have suggested different mechanisms for low temperature and high temperature decomposition of NTO. At lower temperatures, two reaction pathways, which involve proton transfer and internal rotation prior to C-NO_2H cleavage, may be predominant. Whereas at high temperatures, the shortest mechanism, which involve four steps and goes through homolysis of C-NO_2 bond should be the dominant path. Yim et al. have reported very recently a theoretical study of unimolecular decomposition of NTO [92]. The various proposed reaction channels were sampled theoretically by simulating a molecule at high temperature in a number of trajectories. Each of these channels was then further examined to locate the energy barrier and transition structure and to verify the previous results. The C-NO_2 homolysis was again found to be the dominant channel at high temperatures. The departing NO_2 could capture a H atom from the NTO ring to form HONO, by either a concerted bond breaking mechanism or by a bimolecular reaction between the NO_2 group and the triazole ring. At lower temperature, the dissociation channels initiated by hydrogen migration should be activated first. The channel with hydrogen migration followed by ring opening and then by HONO loss has an energy barrier of 159 kJ mol^{-1}, at the rate determining step and it is the path requiring lowest energy among the various paths that have been studied. This study has considered the path involving nitro-nitrite rearrangement [17] also, and was found to have a lower energy barrier than that for C-NO_2 homolysis, but it would make only a minor contribution, due to the entropy factor.

A theoretical study of initial decomposition process of NTO dimer has also been reported very recently [93]. It has been found that there is a reaction path for the production of CO_2 through the dimer reaction with a potential energy barrier of 367.4 kJ mol^{-1}. This study has revealed a series of NTO dimer reaction paths through 4 intermediates and 5 transition states, with the evolution of nitro-TO, CO_2, N_2, HONO and HCN in that order. Thus the study concluded that CO_2 might be produced through a cluster of NTO in the gas phase.

7.9. Explosive Properties of NTO Salts and Their Applications

Safety aspects of high energy materials are one of the major concerns of research activities in this field. The military services worldwide are also adopting policies to make their munitions insensitive (invulnerable) to accidental hazardous stimulii. Thus, compounds with high energy combined with less sensitivity are sought for. Fortunately, NTO is such a compound and is now widely used in explosive formulations, plastic bonded explosives (PBX), gas generator for automobile inflatable air bags systems etc. Since NTO is acidic and thus the ease of formation of salts offer the chance of making a large variety of compounds which can be tailored for a wide spectrum of applications. Although the preparation and properties of a large number of NTO salts are available in open literature, the reports on the applications of these materials are either hidden in classified reports or meagerly available.

Initially ANTO and ethyleneaminoguanidine salts were considered for explosives [94]. However these studies concluded that neither these salts nor those from similar non-energetic amines are likely to show any properties of primary explosives and the potential use of such materials is in low melting eutectic explosives with ammonium nitrate (AN) [95]. ANTO has also been considered as one of the ingredients of composite explosives and found to be having more advantages than AN and AP (ammonium perchlorate) [96, 97]. Redman et al. wanted to use their prepared metal salts as primary explosives, but their screening showed that they have no application as primary explosives [39].

Later, studies on the NTO salts have focused attentions towards the applications in propellants. Lee et al. have assessed the applicability of seven amine salts of NTO as gun propellant ingredients [53, 98]. Their study shows that the compounds have less flame temperature than TAGuN, but with lesser impetus also, when considered as monopropellants. However the mixtures of these salts can be well tailored with HMX, so that to achieve considerably lower flame temperature than pure HMX, with a small compromise with impetus. ANTO has got higher heat of formation (-276.56 kJ mol^{-1}) than the conventional solid propellant ingredients like AP (-296.0 kJ mol^{-1}) and AN (-365.04 kJ mol^{-1}). Given the facts that its combustion is environment friendly than AP and its thermolysis is autocatalytic in nature, it can be considered as a replacement for AP in solid rocket propellants.

The metal salts of NTO have also been suggested to be potential high energetic ballistic modifiers for solid propellants [99]. Williams et al. after their pyrolysis studies concluded that NTO forms a polymeric azine-like residue when it is decomposed below about 280 ^{0}C [21]. This residue can accumulate transiently on the burning surface and retard the mass transfer from condensed phase to gas phase and affect the heat transfer from gas phase to condensed phase. Thus they have hypothetically suggested that NTO can be used as a potential additive in solid propellants for suppressing the burn rate and enhance the combustion stability. Later on the pyrolytic characteristics of NTO and some of its salts have been used as a criterion to evaluate the catalytic activity, when these compounds are used as ballistic modifiers [77]. Singh et al. have also studied the catalytic activity of NTO and its salts as burning rate catalysts for hydroxyl terminated polybutadiene (HTPB)-AP composite solid propellants and found that transition metal salts of NTO have a high potential for this application [78, 100]. Some of the transition metal salts were also evaluated as ballistic modifiers in HTPB-AN composite solid propellants also [101]. The NTO salts alter the temperature for phase transition of NTO. Kulkarni et al. have also explored the metal salts of NTO as energetic ballistic modifiers for solid propellants [102, 103].

The metal salts of NTO need particular attention as high energetic ballistic modifiers for solid propellants. These compounds have a lot of added advantages than the conventional transition metal oxides (TMOs), for the use as burning rate catalysts. TMOs are nonenergetic and hence, the increase of their percentage in propellant formulations reduces total energy of the propellants. The metal salts of NTO can be formulated to be used at higher amounts than that for TMOs with little compromise in energy. The higher catalytic activity of the high energetic salts may also be attributed to the fact that the 'active metal oxides' are formed *insitu* in the system. Brill et al. have shown that the metal salts of NTO forms volatile metal compounds which may be further converted to metal oxides in the gas phase [104]. They have hypothetically suggested the tailoring of these salts so as to dampen the high frequency acoustic modes inside the combustor and thus, avoid unsteady combustion. This is also an added advantage of metal salts of NTO over the conventional TMOs as ballistic modifiers.

However, much work is needed to be carried out on the effect of these salts on the processing parameters as well as ageing of solid propellants, before its actual usage. The hypothetical suggestions also need further experimental support. The energy computations also need experimental assessment to verify the effect of these catalysts on the total energy as well as specific impulse of the propellants.

7.10. Thermolysis of NTO Salts

Thermal behaviour is perhaps one of the most widely studied aspects of the salts of NTO, yet it seems to be insufficient in detail as compared to that of pure NTO. One Ph.D. thesis has been known to be produced on this aspect of NTO salts [105]. Thermoloysis of salts of NTO have usually been undertaken using TG, TG-DTG, DSC/DTA etc. and the mechanistic pathways have been suggested by following the reaction by analysing the intermediate products and residues by using IR and/or XRD. Data from the thermal analysis of some of the metal salts has been summarised in table 7. In the case of hydrated metal salts, the general mechanism of thermolysis may be represented as shown in Scheme 6.

$$M(NTO)_n.mH_2O \xrightarrow[-m\,H_2O]{} M(NTO)_n \xrightarrow{} M\ oxide$$

Scheme 6. Generalised mechanistic pathway for thermolysis of hydrates metal salts of NTO.

where M is metal. For alkali and alkaline earth metal salts the corresponding metal carbonate is the final product of thermolysis.

The dehydration may take place in a single step or multi steps. In the case of alkali metal salts of NTO, the LiNTO directly gives lithium carbonate as the final product of thermolysis, whereas sodium and potassium salts do so via the intermediate formation of corresponding metal cyanate and cyanide respectively [40]. Brill et al. have carried out flash pyrolysis study using T-jump / FTIR spectroscopy and concluded that the intermediate product formed during the pyrolysis of all the alkali metal salts of NTO is metal isocyanate (MNCO) rather than metal cyanate (MCNO) [104].

Among the alkaline earth metal salts MgNTO gives MgO as the end product of thermolysis whereas all the other three salts namely Ca, Sr and Ba forms the corresponding metal carbonates. All the four salts after dehydration undergo NTO ring cleavage to form a mixture of the corresponding metal carbonate, polyamides and volatile substances [41]. However the above said conclusions were drawn from TG and DSC studies in different atmospheres, (i.e. TG was run under N_2 atmosphere whereas DSC under static air atmosphere) yet the data from both techniques is corroborative. Singh et al. undertook thermal studies on the above said alkaline earth metal salts using TG and DTA (both under static air atmosphere) and found that the final product in all cases is metal oxide [106]. They have conducted explosion delay studies (rapid thermolysis) also on these compounds and suggested the mechanism for thermolysis as given in Scheme 7.

The thermolysis of transition metal salts of NTO has been studied in detail and the general mechanism of thermolysis is almost same as shown in Scheme 6 [42, 33,107]. In the case of salts which do not contain water of crystallisation, the first endothermic dehydration process is absent, and thus they undergo denitrification along with NTO ring cleavage and finally the corresponding metal oxides are formed. The silver salt ultimately gives metallic silver due to the instability of its metal oxide [107].

The Cu and Pb salts of NTO are reported to be having two stage exothermic decomposition processes [33]. In the first stage, denitrification takes place along with NTO ring cleavage and thus a mixture containing corresponding metal carbonate is formed. The residue on further heating gives metal oxide which is indicated by a second exothermic peak in DSC. Flash pyrolysis studies were also conducted on Cu and Pb salts of NTO [104]. It shows that the lead salt liberates one or more volatile metal complexes, which may include one or more $Pb(NCO)_2$ isomers and lead cyanide. The presence of co-ordinated water in the gaseous metal salts and polynuclear species also being suspected. But the Cu salt exploded violently and showed the presence of only a small amount of volatile isocyanate like complexes. The primary metal containing species were solid copper oxides. The thermal decomposition of rare earth metal salts [34,43] as well as the metal complexes [45-51] also takes place according to the Scheme 6.

Table 7. Data from thermal analysis and kinetic parameters for NTO and its salts

S.No.	Compound	Peak temperature / ^0C		E_a / kJ mol^{-1}		Log(A)		Ref
		Endo	*Exo*	E_{aI}	E_{aII}	I	II	
1.	NTO		275	520.30		48.10		[33]
2.	$KNTO.H_2O$		230, 257	192.40	118.00	17.10	8.70	[33]
3.	$Mg(NTO)_2.8H_2O$	118, 164	263					[106]
4.	$Ca(NTO)_2.4H_2O$	93, 175	231					[106]
5.	$Sr(NTO)_2.5H_2O$	92	302					[106]
6.	$Ba(NTO)_2.3H_2O$	142, 232	243, 314					[106]
7.	$Mn(NTO)_2$	80, 127	186, 251					[42]
8.	$Fe(NTO)_2$	91	367					[42]
9.	$Fe(NTO)_3.2H_2O$	60	175	115.70	76.77	3.94	2.19	[107]
10.	$Co(NTO)_2$		280					[107]
11.	$Ni(NTO)_2$		312					[42]
12	$Cu(NTO)_2$		355					[42]
13.	$Cu(NTO)_2.4H_2O$		285, 353	115.50	87.10	8.12	6.98	[33]
14.	$Zn(NTO)_2$		346					[42]
15.	$Ag(NTO).H_2O$	130	150, 238, 280	200.60	51.06	6.35	1.11	[107]
16.	$Cd(NTO)_2.2H_2O$	50	257	66.30	41.81	1.91	0.65	[107]
17.	$Hg(NTO)_2.2H_2O$	55, 116	198	129.20	36.67	4.55	0.77	[107]
18.	$Cr(NTO)_3.8H_2O$	68	246	119.00	26.73	4.25	2.19	[107]
19.	$Pr(NTO)_3.9H_2O$		263	153.90		13.10		[43]
20.	$Nd(NTO)_3.7H_2O$		255	130.50		10.90		[43]
21.	$Sm(NTO)_3.8H_2O$		259	148.80		12.70		[43]

In Ref [33] E_{aI} and E_{aII} correspond to the two exothermic decomposition processes, in Ref [107] E_{aI} correspond to endothermic dehydration and E_{aII} correspond to exothermic decomposition process. The peak temperature values in Ref [106] and [107] are from DTA and all other peak values are from DSC. The DSC as well as DTA experiment were carried out in static air atmosphere and heating rate was 10 ^0C min^{-1} except in Ref [106] where it was 15 ^0C $min.^{-1}$.

The ethylene diamine salt of NTO has only one exothermic peak in DSC, which is similar to that of NTO, but the peak temperature is lower than that of NTO [33]. However ANTO has a unique mechanism of thermolysis and is autocatalytic in nature [33]. In a flowing air atmosphere, dehydration is the first process for ANTO followed by deamination and thus NTO is formed. And in the third stage, the residue formed after deamination, goes on decomposing as does NTO. Under a static air atmosphere, the gaseous NH_3 formed by the decomposition of ANTO, catalyses the decomposition of NTO. In the case of RSANTOates thermolysis of some of the salts in static air atmosphere take place in a single exothermic step [108]. But in rest of the salts, the thermal decomposition proceeds via proton transfer (N-H bond heterolysis) to form corresponding amine and NTO molecules [108]. The residue then continues to decompose similar to the decomposition of NTO. But any rationalisation for this behaviour is not available. The mechanistic pathway suggested for thermolysis of RSANTOates is given in Scheme 8.

7.10.1. Kinetics of Thermolysis of NTO Salts

Kinetics of thermolysis is also usually presented along with mechanistic aspects for many salts of NTO and the values of activation energy (E) and pre-exponential factor (log A) are summarised in table 7. However, these data are inadequate in the sense that an isothermal kinetic analysis is seldom reported. The kinetic parameters reported are usually evaluated by using non-isothermal methods of Ozawa [109], Kissinger [110], Coats-Redfern [111] etc. The values of apparent activation energy (E) for the thermolysis of NTO, determined by the same method differ considerably (365.8 [33], 520.3 $kJmol^{-1}$ [112]). However the values of E for the salts are more consistent. A direct comparison of thermal stability of the entire range of NTO salts on the basis of values of E summarised in table 7 is not possible. This is because the kinetic parameters may suffer from the complexities imposed by the conditions under which the values were obtained. However a legitimate comparison is possible as far as the conditions are same. Thus on the basis of E, the thermal stability of NTO and its salts follows the order NTO >> PbNTO > ENTO > KNTO > CuNTO [33] and NTO > FeNTO > AgNTO > CdNTO > HgNTO > CrNTO [107]. Thus it can be seen that the thermal stability of NTO is lowered when it gets converted into its salts. This may be due to the catalytic effect of the metals on the thermolysis of NTO, in their corresponding metal salts.

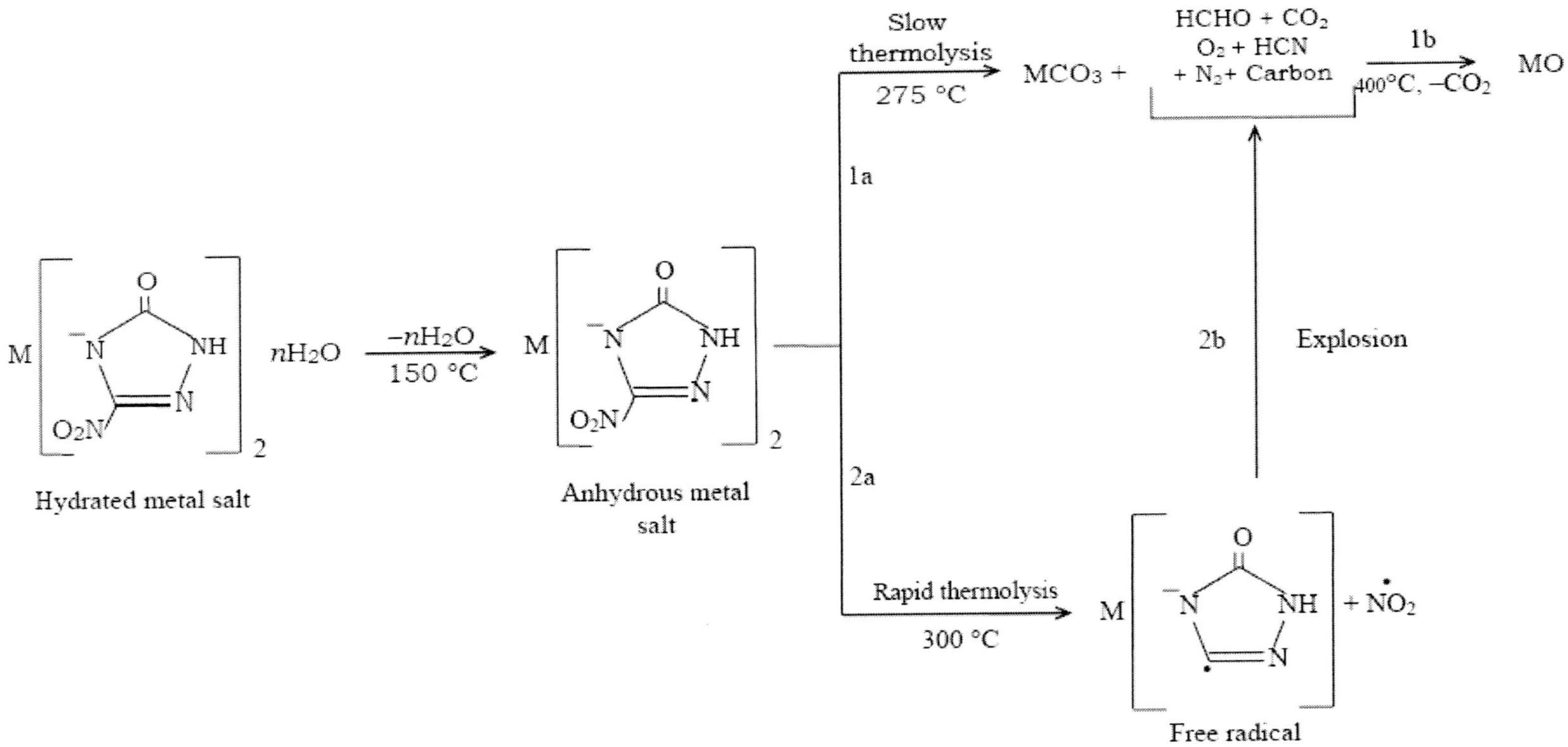

where $n = 3 - 8$

Scheme 7. Schematic representation of thermolysis of alkaline earth metal salts of NTO.

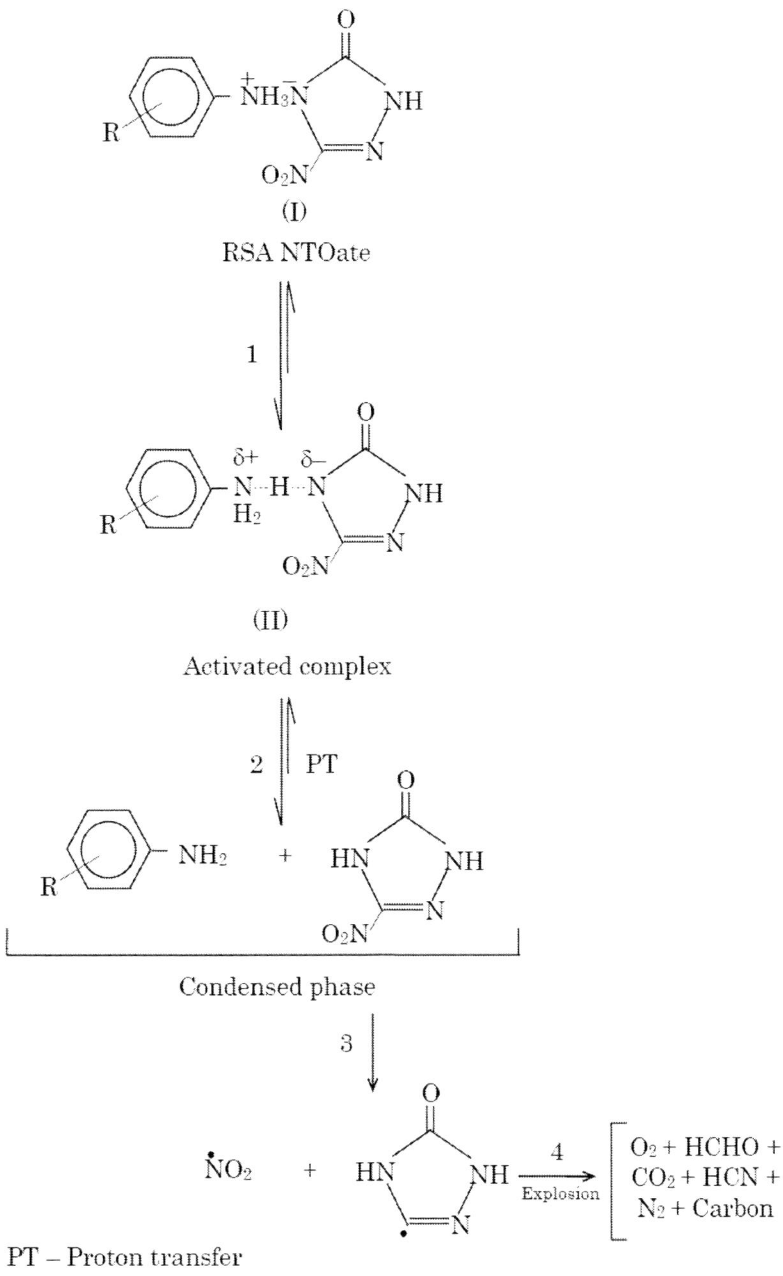

Scheme 8. Schematic representation of thermolysis of RSANTOates.

7.10.2. Thermodynamic, Thermochemical and Performance Parameters of Salts of NTO

Thermodynamic and thermochemical parameters such as heat of formation (ΔH_f), lattice energy, lattice enthalpy etc. are of great importance for any energetic compound. Performance parameters such as impetus, flame temperature, sensitivity to impact, friction, spark etc. and critical temperature for thermal explosion (T_b) are also important from the application point of view. The details of these parameters for the salts of NTO are briefly discussed below.

7.10.2.1. Thermodynamic and Thermochemical Properties

The standard enthalpies of formation of silver, ammonium, sodium and pottassium salts of NTO have been determined by Finch et al. [113, 114]. Later on the Chinese scientists have determined standard enthalpies of formation ($-\Delta_f H^\theta_m$) of KNTO.H_2O, Ba$(NTO)_2$.3H_2O, LiNTO.2H_2O, Ca$(NTO)_2$.4H_2O and Gd$(NTO)_3$.7H_2O from the measurements of enthalpy of solution in water using a Calvet micro-calorimeter [115, 116]. From measurements of enthalpy of precipitation of KNTO.H_2O crystal with $Pb(NO_3)_{2(aq)}$, $CuSO_{4(aq)}$ and $Zn(NO_3)_{2(aq)}$, the standard enthalpy of formation of Pb$(NTO)_2$.H_2O, Cu$(NTO)_2$.2H_2O and Zn$(NTO)_2$.H_2O were also determined and the values are summarised in table 8. The standard enthalpies of formation of fourteen other metal salt hydrates are also reported to be calculated from their respective enthalpies of dissolution in water and the data is given in table 8 [117]. With the help of all above mentioned data, literature data and Kapustinskii's equation [118], the lattice energies (ΔU^θ_l) and lattice enthalpies (ΔH^θ_l) of twenty metal salts of NTO were determined. The lattice energies, lattice enthalpies and standard enthalpies of dehydration (ΔH^θ_d) were also estimated for the above twenty metal salts and are given table 8. The above said thermodynamic as well as thermochemical parameters are reported for the metal complexes of NTO also [45-51]

Table 8.The standard enthalpy of formation, lattice enthalpy, lattice energy and standard enthalpy of dehydration for metal salts of NTO

Compound	$-\Delta_f H^\theta_m$/ (kJ mol^{-1})	$-\Delta H^\theta_l$ / (kJ mol^{-1})	$-\Delta U^\theta_l$/(kJ mol^{-1})	ΔH^θ_d/(kJ mol^{-1})
Li(NTO).2H_2O	966.6 ± 2.2	788.13	778.21	6.21
Na(NTO).H_2O	650.6 ± 2.4	643.36	635.92	2.17
K(NTO).H_2O	676.9 ± 2.6	574.98	567.54	5.97
Mg$(NTO)_2$.8H_2O	3002.6 ± 4.5	2667.74	2640.46	269.86
Ca$(NTO)_2$.4H_2O	1905.5 ± 4.4	2115.32	2097.96	70.48
Mn$(NTO)_2$.8H_2O	2759.7 ± 4.7	2595.54	2568.26	279.66
Co$(NTO)_2$.8H_2O	2594.7 ±4.5	2752.94	2725.66	211.06
Cu$(NTO)_2$.2H_2O	712.1 ± 5.4	2533.66	2521.26	438.84
Zn$(NTO)_2$.2H_2O	628.8 ± 5.7	2420.88	2410.96	381.07
Y$(NTO)_3$.6H_2O	2755.2 ± 6.7	4395.08	4370.28	340.32
La$(NTO)_3$.7H_2O	3031.8 ± 6.6	4120.36	4093.08	145.59
Ce$(NTO)_3$.7H_2O	3020.9 ± 6.5	4168.86	4141.58	169.09
Pr$(NTO)_3$.7H_2O	3025.6 ± 6.7	4211.66	4184.38	192.89
Nd$(NTO)_3$.8H_2O	3304.9 ± 6.7	4288.44	4258.68	208.66
Sm$(NTO)_3$.7H_2O	3025.0 ± 6.5	4304.36	4277.08	238.59
Eu$(NTO)_3$.7H_2O	2936.8 ± 6.6	4352.06	4324.78	271.99
Gd$(NTO)_3$.7H_2O	3020.1 ± 6.4	4370.06	4342.78	278.99
Tb$(NTO)_3$.5H_2O	2437.0 ± 6.5	4302.00	4289.68	283.95
Dy$(NTO)_3$.5H_2O	2452.2 ± 6.5	4326.80	4314.48	293.75
Yb$(NTO)_3$.6H_2O	2708.6 ± 6.6	4516.38	4491.58	389.62

The ΔH_f of ENTO, ANTO, HNTO, GuNTO, AGuNTO, DAGuNTO and TAGuNTO are calculated from the values of heat of combustion measured using a parr bomb calorimeter. Based on the measured ΔH_f, values for impetus and flame temperature of these salts as monopropellants were also calculated [53]. A more positive value for ΔH_f was obtained as the nitrogen content of the guanidium homolog salt increases.

7.10.2.2. Explosive Properties of Salts of NTO

Extensive measurements of explosion delay on NTO as well as the metal salts [42, 106, 107] and RSANTOates [108] have been reported. The activation energy for thermal explosion has also been determined and it shows that the metal salts are thermaly less stable than NTO, whereas the RSANTOates showed almost similar thermal stability as that of NTO.

Redman et al. after their studies on salts of NTO using 'Ball and disc' method, concluded that the response of the materials is typical of sensitive high explosives [39]. However Singh et al. have assessed the impact as well as friction sensitivity of ten transition metal salts of NTO and found that all these salts may be classified as insensitive explosives [119]. The spark sensitivity has also been measured and was found to be insensitive upto 5 J of electrical energy. The small scale sensitivity studies on the amine salts of NTO show that the compounds are significantly less sensitive to impact and spark than the common propellant ingradients such as RDX, triamino guanidium nitrate (TAGuN) etc [53].

The value of T_b for NTO has been reported by Lee et al. [6]. Later on T_b for NTO and its salts (ANTO, ENTO, CuNTO and PbNTO) has been determined by the Chinese scientists using four different methods and the values are summarised in table 9 [112]. The T_b values were experimentally obtained and the values obtained are compared with that obtained from three other theoretical methods (Frank-Kamenetski method [118], Zinn-Mader-Rogers method [120, 121] and a calculation formula under non-isothermal DSC condition [122]). The values of T_b obtained using all the four methods are found to be within the limit of experimental error.

Table 9. Experimental and theroretical values of critical temperature of thermal explosion (T_b) for NTO and its salts

Method	NTO	ENTO	KNTO	PbNTO	CuNTO	HMX	RDX	PETN	Tetryl
T_b [K]	513.0	497.0	491.0	472.0	463.0	529.0	488.0	478.0	473.0
$T_{b(F\text{-}K)}$[K]	520.8	488.8	495.2	454.0	475.2	--	--	--	--
$T_{b(Z\text{-}M\text{-}R)}$ [K]	529.6	517.5	495.6	479.6	--	--	--	--	--
$T_{b(DSC)}$ [K]	528.2	493.3	488.3	464.3	445.3	--	--	--	--

$T_{b(F\text{-}K)}$[K], $T_{b(Z\text{-}M\text{-}R)}$ [K], $T_{b(DSC)}$ [K] are the values of T_b obtained by Frank-Kamenetskii's method, Zinn-Mader-Roger's method and non-isothermal DSC method respectively.

The thermal stability of NTO as well as its salts have been compared with that of RDX, HMX, tetryl and PETN on the basis of values of T_b and the order is as follows:

HMX > NTO > ENTO > KNTO > RDX > PETN > tetryl > PbNTO > CuNTO.

7.11. Conclusion

The lack of consensus in the thermolysis mechanism of NTO is due to the dependence of the mechanism on the experimental configurations and sample characteristics. Thermolysis of NTO is complex and the mechanism varies according to the conditions of the experiment. Theoretical studies on the thermolysis of NTO also confirmed this apprehension. There exists many decomposition pathways with minor differences in energy and thus feasible under different conditions. It is very difficult to ascertain a single mechanism for decomposition of NTO. Most of the experiments gave CO_2 as the initial product gas, which can only be explained by bimolecular routes. However almost all the theoretical studies proved that C-NO_2 homolysis is the initial step, that requires lowest energy. This is due to the fact that most of the theoretical studies considered unimolecular decomposition pathways. Studies concerning bimolecular routes or NTO clusters indeed show the feasibility of CO_2 production. Thus, as the experimental studies have been made on bulk samples, the interaction between NTO molecules is possible and CO_2 is produced initially rather than NO_2. The sensitivity data available on NTO proves the insensitivity of this compound. Performance parameters also show the potential of NTO for various applications. Thus NTO could be used for applications, which requires insensitivity and high performance.

NTO forms salts with a wide range of compounds and the salts formed may be generally classified as metal salts and amine salts. The proton leaves from position N(4) when NTO is converted to its salts. Generally the metal salt hydrates of NTO during their thermolysis first undergo dehydration. Then the dehydrated salts on further heating undergo exothermic decomposition ultimately yielding the corresponding metal oxides. Thermal stability of almost all the metal salts is lower than that of NTO, on account of values of E for thermolysis and critical temperature for thermal explosion. All salts of NTO are also rather insensitive like the parent compound. The metal salts of NTO are potential energetic ballistic additives for solid propellants, but further studies are required.

References

[56] V.W. Manchot, R.Noll, *Justus Liebigs Annalen der Chemie.* 343 (1905) 1.

[57] H. Gehlen, J. Schmidt, *Justus Liebigs Annalen der Chemie.* 682 (1965) 123.

[58] G.I. Chipen, R.P. Bokalder, V.Y. Grinshtein, *Chem. Heterocycl.* 2 (1966) 79.

[59] G.I. Chipen, R.P. Bokalder, V.Y. Grinshtein, *Khim. Geterotsikl. Soedin.* 2(1) (1966) 110.

[60] C.F. Kroger, R. Miethchen, H. Frank, M. Siemer, S. Pilz, *Chemische Berichte.* 102 (1969) 755.

[61] K.Y. Lee, L.B. Chapman, M.D. Coburn, *J. of Energ. Mater.* 5 (1987) 27.

[62] G. Singh, I.P.S. Kapoor, S.K. Tiwari, P.S. Felix, *J. Hazard. Mater.* B 81 (2001) 67.

[63] G. Singh, I.P.S. Kapoor, S.M. Mannan, S.K. Tiwari, *J. Hazard. Mater.* A 68 (1999) 155.

[64] G. Singh, S. Prem Felix, *J. Hazard. Mater.* A90 (2002) 1.

[65] G. Singh, S. Prem Felix, *J. Molec. Struct.* 649 (2003) 71.

[66] H. Ostmark, H. Bergman, G. Aquist, *Thermochim. Acta.* 213 (1993) 165.

[67] K.V. Prabhakaran, S.R. Naidu, E.M. Kurian, *Thermochim. Acta.* 241 (1994) 199.
[68] D.C. Sorescu, D.L. Thompson, *J. Phys. Chem.* 101 (1997) 3605.
[69] L. Jiarong, C. Boren, O. Yuxiang, *Propellants Explos. Pyrotech.* 16 (1991) 145.
[70] J.A. Menapace, J.E. Martin, D.R. Bruss, R.V. Dascher, *J. Phys. Chem.* 95 (1991) 5509.
[71] J.C. Oxley, J.L. Smith, E. Rogers, X.X. Dong, *J. Phys. Chem.* A 101 (1997) 3531.
[72] N.L. Garland, H.D. Ladouceur, H.H. Nelson, *J. Phys. Chem.* 101 (1997) 8508.
[73] G.K. Williams, T.B. Brill, *J. Phys. Chem.* 99 (1995) 12536.
[74] J.C. Oxley, J.L. Smith, Z. Zhou, *J. Phys. Chem.* 99 (1995) 10383.
[75] T.B. Brill, P.E. Gongwer, G.K. Williams, *J. Phys. Chem.* 98 (1994) 12242.
[76] G.K. Williams, S.F. Palopoli, T.B. Brill, *Combust. Flame.* 98 (1994) 197.
[77] D.F. McMillan, D.C. Erlich, C. He, C.H. Becker, D.A. Shockey, *Combust. Flame* 111(3) (1997) 133.
[78] H. Ostmark, H. Bergman, G. Aquist, A. Langlet, B. Persson, *Proceeding of the 16th International Pyrotechnics Seminar,* 1991 p. 874.
[79] H. Ostmark, FOA report D-201782.3, National Defence Research Establishment, Sundbyburg, Sweden, Nov. 1991.
[80] G.T. Long, B.A. Brems, C. A. Wight, *J. Phys. Chem.* B 106 (2002) 4022.
[81] E.F. Rothgery, D.E. Audette, R.C. Wedlich, D.A. Csejka, *Thermochim. Acta.* 185 (1991) 235.
[82] T.B. Brill, P.J. Brush, D.G. Patil, *Combust. Flame* 92 (1993) 178.
[83] K.Y. Lee R. Gilardi, in: Structure and Properties of Energetic Materials, Research Society, Pittsburgh, PA Vol. 296, 1993, pp. 237.
[84] J.P. Ritchie, *J. Org. Chem.* 54 (1989) 3553.
[85] A. Finch, P.J. Gardner, A.J. Head H.S. Majdi, *J. Chem. Thermodyn.* 23 (1991) 1169.
[86] A. J. Sanderson, A Draft Data Sheet of 3-nitro-1,2,4-triazole-3-one, NIMIC Database I: The NIMIC Co-ordinated Characterisation Programme, 1997.
[87] E. A. Zhurova, A. A. Pinkerton, *Acta Cryst.* B57 (2001) 359.
[88] X. Yi, H. Rongzu, W. Xiyou, F. Xiayun, Z. Chunhua, *Thermochim. Acta.* 189 (1991) 283.
[89] Z. Tonglai, H. Rongzu, L. Fuping, in: *Proc. 3rd Int. Symp. Pyrotech. Explos.,* Beijing, 1995 pp. 303 – 306.
[90] Z. Tonglai, H. Rongzu, L. Fuping, K. Yu, *Acta Chim. Sin. 52* (1994) 545.
[91] Z. Tonglai, H. Rongzu, L. Fuping, L. Chen, K. Yu, *Chin. Sci. Bull.* 38 (1993) 1350.
[92] Z. Tonglai, H. Rongzu and L. Fuping, in: Proc. 9th *Symp. Chem. Problems Connected With the Stability of Explos.,* Margretertorp, Sweden, 1993 pp. 259 – 265.
[93] Z. Tonglai, H. Rongzu, L. Fuping, *Energ. Mater.* 1(4) (1993) 827.
[94] L.D. Redman, R.J. Spear, MRL Technical Note, An Evaluation of Metal Salts of NTO as Potential Primary Explosives, MRL-TN-563 1989.
[95] Z. Tonglai, H. Rongzu, L. Yanjun, L. Fuping, *J. Therm. Anal.* 39 (1993) 827.
[96] Z. Tonglai, H. Rongzu, L. Fuping, *Thermochim. Acta.* 244 (1984) 185.
[97] G. Singh, I.P.S. Kapoor, S.K. Tiwari, S.P. Felix, *Indian J. of Engg. Mater. Sci.* 7 (2000) 229.
[98] X. Yi, H. Rongzu, Z. Tonglai, L. Fuping, *J. Therm. Anal.* 39 (1993) 41.
[99] Z. Tonglai, H. Rongzu, L. Fuping, *J. of Rare Earths.* 13(1) (1995) 10.
[100] H. Rongzu, S. Jirong, L. Fuping, K. Bing, K. Yanghui, M. Zhihua, Z. Zonghua, H. Zhou, *Thermochim. Acta.* 299(1-2) (1997) 87.

[101] S. Jirong, H. Rongzu, K. Bing, L. Fuping, *Thermochim. Acta.* 331(1) (1999) 49.

[102] S. Jirong, H. Rongzu, K. Bing, X Yi, L. Fuping, K. Yu, *J. Therm. Anal. Calorim.* 55(3) (1999) 797.

[103] S. Jirong, H. Rongzu, L. Fuping, Z. Tonglai, M. Zhihua, Z. Zonghua, H. Zhou, *Chin. Sci. Bull.* 41 (1996) 1806.

[104] S. Jirong, H. Rongzu, K. Bing, L. Fuping, *Thermochim. Acta.* 335 (1999) 19.

[105] S. S. Yun, J. K. Kim, C. H. Kim, *J. Alloys and Compounds,* 408-412 (2006) 945.

[106] S. Jirong, N. Binke, H. Rongzu, K. Bing, *Thermochim. Acta.* 352-353 (2000) 111.

[107] K.Y. Lee, M.D. Coburn, *US Patent* 4,733,610 1988.

[108] K.Y. Lee, M.M. Stincipher, Propellants, *Explos., Pyrotechnics.* 14 (1989) 241.

[109] G. Singh, I.P.S. Kapoor, S.M. Mannan, S.K. Tiwari, *J. Energ. Mater.* 16 (1998) 31.

[110] D.T. Cromer, J.H. Hall, K.Y. Lee, R.R. Ryan, *Acta Crytallogr. Sect. C: cryst. Struct. Commun.* 44 (1988) 1144.

[111] J.P. Ritchie, K.Y. Lee, D.T. Cromer, E.M. Crober, D.D. Lee, *J. Org. Chem.* 55:7 (1990) 1994.

[112] C.W. Chang, Y.M. Wong, T.C. Chang, C. Chen, *Propellants, Explos., Pyrotech.* 22(4) (1997) 240.

[113] Z. Tonglai, H. Rongzu, L. Fuping, K. Yu, *Energ. Mater.* 1 (1993) 37.

[114] A. Becuwe, Fr. Demande FR 2584006 A1, (1985).

[115] A. Becuwe, Fr. Demande FR 2584068 (1987).

[116] A. Becuwe and A. Declos, *ICT Int. Jahrenstag, Karlstruhe, FRG,* 1987, paper 27.

[117] A. Becuwe and A. Declos, Proc. 9th Symp. On Detonation, Portland, OR, US Department of Energy, 1989, p. 871.

[118] S. Cudzilo, W.A. Trzcinski, *J. of Energ. Mat.,* 19 (2001) 1.

[119] F. Bugaut, 'Performances des Nouveaux Explosifs Insensible, Cas du NTO', Deuximes Journees Scientifiques Paul Ville, Brest, 1991.

[120] M. Piteau, A. Becuwe and B. Finchi, 'New Energetic Ingredients for Insensitive Munitions and Clean propellants', ADPA Symp., San Diego, California, 1991.

[121] J.P.Agrawal, S.M. Walley and J.E. Field, *Combust. Flame,* 112 (1998) 62.

[122] R.J. Spear, C.N. Louey and M.G. Wolfson, Materials Research laboratory, DSTO, Report MRL-TR-89-18 (1989).

[123] C.C. Hung and T.S. Wu, *Thermochim. Acta,* 239 (1994) 105.

[124] F. Volk and H. Bathelt, in *10th Symp. Chem. Probl. Connected Stab. Explos., Fraunhofer Institute of Chemical Technology, Pfiztal,* Germany, 1996 p. 319.

[125] F.C. Fouche, van Schalkwyk and C. Gideon, in *Proc. 27th Int. Annu. Conf. ICT, Pfinztal,* Germany, 1996, p. 69.1.

[126] Y.H. Kong, Z.R. Lin, Y.H. Shao, C.M. Yin and W. He, *Thermochim. Acta,* 297(1-2) (1997) 1.

[127] T.S. Sunrall, in *Proc. 23rd Int. Pyrotech. Semin., IIT Research Institute,* 1997, p. 822.

[128] T.S. Sunrall, in *Proc. 23rd Int. Pyrotech. Semin. IIT Research Institute,* 1997, p. 847.

[129] G.J. Ellis and H.C. Benzuidenhout, in *Proc. 30th Int. Annu Conf. ICT, Pfinztal,* Germany, 1999, 63.1.

[130] F.C. Fouche, H.C. Benzuidenhout, C.E. Du Toit, F.A. Venter and J.H. Niemand, *Proc 30th Int. Annu. Conf. ICT,* 1999, p. 48/1.

[131] T. Mukundan, G. N. Purandre, J. K. Nair, S. M. Pansare, R. K. Sinha, H. Singh, *Defence Sci. J.* 52 (2002) 127.

[132] G. Dalin, L. Shangwen, Y. Cuimei, Z. Shuangjian and D. Yongzhan, Tuijin Jishu, 20(3) (1999) 91.
[133] G. Singh and S.P. Felix, *Combust. Flame,* 132 (2003) 422.
[134] E. Gast, B. Schmid and P. Semmler, *PCT Int. Appl.* WO 994843 A1 (1999).
[135] J.B. Canterberry, S.S. Shlueter, J.H. Adams and R.K. Walsh, *PCT Int. Appl. WO* 9908983 A1 (1999) 110.
[136] K.A.Wilson, M.A.Wilson, D.R. Poole and G.F. Holland, US 5866842 A2 (1995) 5.
[137] C.B. Storm, J.R. Stine and J.F. Kramer, *Sensitivity Relationship in Energetic Materials – Chemistry and Physics of Energetic Materials, Kluwer Academic Publishers,* Dordrecht, 1990, p. 605.
[138] B.C. Beard, J. Sharma, *J. Energ. Mater.,* 7(3) (1990) 181.
[139] B.C. Beard, J. Sharma, *J. Energ. Mater.,* 11 (1993) 325.
[140] J.C. Oxley, J.L. Smith, K.E. Yeager, E. Rogers and X.X. Dong, *Proc. Mater. Res. Soc. Symp.* 418, 1996, p. 135.
[141] D.J. Beardall, T.R. Botcher and C.A. Wight, *Proc. Mat. Res. Symp.,* 418, 1996, p. 379.
[142] T.R. Botcher, D.J. Beardall, C.A. Wight, L. Fan and T.J. Burkey, *J. Phys. Chem.,* 100 (1996) 8802.
[143] M.R. Ottaway, R.J. Granville, *Proc. 8th Symp. Chem. Probl. Connetced Stab. Explos. Columbia Sci. Ind., Milton Keynes, UK,* 1998, p. 1.
[144] N. J. Harris and K. Lammertsma, *J. Amer. Chem. Soc.,* 34 (1996) 8048.
[145] C. Meredith, T.P. Russell, R.C. Mowrey, J.R. Mc Donald, *J. Phys. Chem. A,* 102(2) (1998) 471.
[146] Y.M. Wang, C. Chen and S.T. Lin, *J. of Mol. Struct.,* 460(1-3) (1999) 79.
[147] Y.W. Leung and L.Z. Feng, *J. Amer. Chem. Soc.,* 123(10) (2001) 2243.
[148] Y. Kohno, O. Takahashi and K. Saito, *Phys. Chem. Chem. Phys.,* 3(14) (2001) 2742.
[149] J.P. Ritchie, E.M. Kober, in: Proc. 9th Int. Symp. On Detonation, Portland, OR, USA, Vol II, 1989 pp. 528.
[150] K.Y. Lee, M.D. Coburn, Report No. LA-10302-MS, Los Alamos National Laboratory, New Mexico, 1985.
[151] T.P. Korkman, *Zh. Organich. Khimii.* 16 (1979) 420.
[152] K.Y. Lee, Report No.LA-UR-84-2038, Los Alamos National Laboratory, Los Alamos, New Mexico, USA.
[153] K.Y. Lee, M.M. Stincipher, *US Patent* 5256792, 1993.
[154] L. Shangwen, W. Jiangming, F. Xiayun, Z. Jihua, E*nerg. Mater.* 1 (1993) 22.
[155] G. Singh, S.K. Tiwari, S. P. Felix, T.L. Varghese, K.N. Ninan, *J. Energ. Mater.,* 20 (2002) 309.
[156] G. Singh, S. P. Felix, *Combust. Flame,* 135 (2003) 422.
[157] P.B. Kulkarni, T.S. Reddy, J.K. Nair, A.N. Nazare, M.B. Talawar, T. Mukundan, S.N. Asthana, Technical Report HEMRL/EMR/TR/2004/4.
[158] P.B. Kulkarni, T. S. Reddy, J.K. Nair, A. N. Nazare, M.B. Talawar, T. Mukundan, S.N. Asthana, *J. Hazard. Mater.* 123 (2005) 54.
[159] T.B. Brill, T.L. Zhang, B.C. Tappan, *Combustion and Flame.* 121 (2000) 662.
[160] S.K. Tiwari, Synthesis and Thermal Decomposition of NTO and its Salts, Ph.D. thesis submitted to DDU Gorakhpur University 2000.
[161] G. Singh, I.P.S. Kapoor, S.K. Tiwari, *Indian J. of Chem. Techn.,* 7 (2000) 236.

[162] G. Singh, I.P.S. Kapoor, S.P. Felix, J.P. Agrawal, *Propellants, Explos., Pyrotech.* 27 (2002) 16.
[163] G. Singh, I.P.S. Kapoor, S.M. Mannan, S.K. Tiwari, *J. Energ. Mater.* 16 (1998) 101.
[164] T. Ozawa, *Bull. Chem. Soc. Japan.* 38 (1965) 1881.
[165] H.E. Kissinger, *Anal. Chem.* 29 (1957) 1702.
[166] A.W. Coats and J.P. Redfern, *Nature. 201*, 68 (1964).
[167] X. Yi, H. Rongzu, Y. Chaoquing, F. Guofu, Z. Jihua, Propellants, *Explos., Pyrotech.* 17 (1992) 298.
[168] A. Finch, P.J. Gardner, A.J. Head, H.S. Majdi, *Thermochim. Acta.* 213 (1993) 17.
[169] A. Finch, P.J. Gardner, A.J. Head, H.S. Majdi, *J. Chem. Thermodyn.* 23(12) (1991) 1169.
[170] M. Zihui, H. Rongzu, *J. of Therm. Anal.* 45 (1995) 79.
[171] H. Rongzu and M. Zihui, in: *Proc. of* 3rd *Beijing Int. Symp. Pyrotech. and Explos., Beijing, China 1995, Beijing Inst. of Techn.* Press, 1995, pp. 160-166.
[172] H. Rongzu, M. Zihui, K. Bing, *Thermochim Acta.* 275 (1996) 159.
[173] D.A. Frank-Kamenetskii, *J. Phys. Chem.* (U.S.S.R.) 13 (1939) 738.
[174] G. Singh, Progress Report of the ISRO RESPOND project No.10/3/313 2001.
[175] J. Zinn, C.L. Mader, *J. Applied Physics.* 31(2) (1960) 323.
[176] J. Zinn, R.N. Rogers, *J. Phys. Chem.* 66(12) (1962) 2646.
[177] Z. Tonglai, H. Rongzu, X. Yi, L. Fuping, *Thermochim. Acta.* 244 (1994) 171.

Chapter 8

THERMOLYSIS OF PLASTIC BONDED EXPLOSIVES

ABSTRACT

A systematic study of thermolysis of plastic bonded explosives of most important explosives compounds, RDX, HMX and keto-RDX is reviewed in this chapter. The results from various thermoanalytical techniques are analysed and discussed by noting the difference between characteristic temperatures for the thermolysis of pure energetic compound and its corresponding PBXs. In general, coating explosives crystals with thermoplastic binders reduces characteristic temperatures for thermolysis. However, sensitivity to hazardous stimuli decreases for the energetic filler in its PBXs. The reason for reduction in thermal stability of energetic compounds in its PBXs may be different for different explosive molecules depending on the nature of the energetic compound and the binder. Kinetic parameters for thermolysis using conventional model fitting approach often suffers from 'kinetic compensation' and may not reflect the complexities of underlying thermochemistry. Use of model-free isoconversion methods gives a better reflection of the complex thermochemistry.

8.1. INTRODUCTION

Plastic bonded explosives (PBXs) are typical composite energetic materials (CEMs) in which a high energetic compound is filled in a polymer (binder) matrix. Coating an energetic compound thoroughly with plasticized thermoplastic polymers can reduce the sensitivity of them to hazardous stimuli such as shock, impact, friction spark etc [1]. The function of the binder is to protect the explosive crystals from quick initiation due to outer stimuli. This is realized by coating the entire crystal surfaces of the explosive by a thin polymer layer. This layer can absorb the accidental energy pulses and distribute them evenly over the volume of the PBX, thereby preventing local decomposition and initiation. Another important function of the binder is its high mechanical strength: the binder will create a void free matrix between the explosive particles. The so formed PBX has a high mechanical strength that prevents crack formation even under high pressure conditions.

Conventional as well as energetic polymers are being used as binders for PBXs. Some of the most common binders are summarized in table 1.

Table 1. Binders used in plastic bonded explosives

Material	Description
Estane 5702-F1	Polyurethane solution system
Viton A	Vinylidene fluoride/hexafluoropropylene copolymer, 60/40 wt %
BDNPA-F	Bis(2,2-dinitropropyl)acetal/bis(2,2-dinitropropyl) formal, 50/50 wt %
CEF	Tris-β-chlorethyphosphate
Kel-F 800	Chlorotrifluoroethylene/vinylidine fluoride copolymer, 3:1
FPC 461	Vinyl chloride/chlorotrifluoroethylene copolymer, 1.5:1
Polystyrene	Styrene polymer, 100%
FEPO	Bis(2-fluoro-2,2-dinitroethyl)formal

The binder is usually precipitated out of solution in the preparation process such that it coats the explosive crystals. Agglomerates of these coated crystals form pressing "beads'. The beads are then either die pressed or isostatically pressed at high temperatures as high as 120 ^{0}C. Pressures from 1 to 20 kPsi then produce billets or pellets, with density as high as 97 % of the theoretical maximum density (TMD). Table 2 lists various PBXs in common use in the U.S. Department of Energy's weapons laboratories.

Table 2. Some typical plastic bonded explosives

Name	Explosive Ingredients	Binder Ingredients	Colour
LX-14-0	HMX, 95.5%	Estane 5702-Fl, 4.5%	Violet spots on white
LX-10-0	HMX, 95%	Viton A, 5%	Blue-green spots on white
LX-10-1	HMX, 94.5%	Viton A, 5.5%	Blue-green spots on white
PBX-9501	HMX, 95%	Estane, 2.5%; BDNPA-F 2.5%	White
PBX-9404	HMX, 94%	NC, 3%; CEF, 3%	White or blue
LX-09-1	HMX, 93.3%	BDNPA, 4.4%; FEPO, 2.3%	Purple
LX-09-0	HMX, 93%	BDNPA, 4.6%; FEFO, 2.4%	Purple
LX-07-2	HMX, 90%	Viton A, 10%	Orange
PBX-9011	HMX, 90%	Estane and 5703-Fl, 10%	Off-white
LX-04-1	HMX, 85%	Viton A, 15%	Yellow
LX-11-0	HMX, 80%	Viton A, 20%	White
LX-15	HNS-1, 95%	Kel-F 800, 5%	Beige
LX-16	PETN, 96%	FPC 461, 4%	White
PBX-9604	RDX, 96%	Kel-F 800, 4%	
PBX-9407	RDX, 94%	FPC461, 6%	Black or white
PBX-9205	RDX, 92%	Polystyrene, 6%, DOP, 2%	White
PBX-9007	RDX, 90%	Polystyrene, 9.1%; DOP, 0.5%; rosin, 0.4%	White or mettled gray
PBX-9010	RDX, 90%	Kel-F 3700, 10%	White
PBX-9502	TATB, 95%	Kel-F 800, 5%	Yellow
LX-17-0	TATB, 92.5%	Kel-F 800, 7.5%	Yellow
PBX-9503	TATB, 80%; HMX, 15%	Kel-F 800, 5%	Purple

Apart from their extensive applications in munitions, PBXs also find applications in flight of launch vehicles. Pyrovalves, cable cutters, explosive latches, explosive zip, flexible linear shaped cords etc. are used for several functions. Pyrovalves are used to connect high-pressure line to downstream element otherwise positively disconnected because of safety requirements. Explosive bolts are used to separate the elements on demand. Flexible linear cords are used for stage separation and also vehicle destruct systems which need to get activated in case the vehicle deviates from the intended path.

Thermal characterization is an essential step in the study of energetic materials. This is because the above stated hazardous stimuli eventually trigger off a thermal event, which is one of the main causes of initiation of any explosive [2]. Thermal decomposition pathways and products influence the performance of any HEM. Thermal characterization and analysis of energetic materials is important, not only from the point of view of safety and application, but also from the academic point of view. The data from thermal characterization and its kinetic analysis may help in the research efforts in this field, by predicting the safety of the future formulations. A systematic study on the thermochemistry of PBXs is not readily found in open literature [3]. Most of the studies reported on PBXs are either routine characterizations [4] or are not aimed at understanding the thermo-chemistry [5,6]. Thus, these studies are insufficient to understand the effect of coating on thermal decomposition mechanism and kinetics of the energetic compound in PBXs. Moreover, studies on nitramine propellants [7-13] that are essentially CEMs and similar in formulation to PBXs, show that the binder plays a significant role during thermolysis. Oyumi [13] showed that the presence of bis(azidomethyl)oxetane / tetrahydrofuran (BAMO), which is a copolymer binder, initiated and accelerated the rate of HMX thermal decomposition in their composite propellant. Recently, Chaves et al. [3] have studied the influence of binder (HTPB), plasticiser (bis 2-ethylhexylsebacate, DOS) and curative (isophorone diisocyanate, IPDI) on the thermal decomposition of PBX, based on RDX. They have concluded that although the thermal stability of RDX does not change in the PBX, the additives change the reaction pathways. Singh et al. [14] have showed that plasticisers have no remarkable effect on thermal characteristics of the PBX (RDX-polyurethane). Thermal characterization of PBX containing HMX and HTPB has been reported by Lee et al. [15]. In the light of above facts, a systematic study was done by G. Singh et al. and some of the results on thermolysis of PBXs containing HMX, RDX and keto-RDX as energetic fillers and Estane, Kel F and Viton A as binders, are discussed in this chapter. The results of these investigations are analysed by noting down the changes in the thermal behaviour of the mixture (PBX) from the thermal behaviour of pure energetic compound studied individually.

There are a number of correlations available in literature between kinetics of thermolysis in HEMs and performance parameters. Cook et al. [16, 17] proposed that the kinetics of initial reactions are important in determining detonation velocity. Zeman et al. correlated thermal decomposition kinetics of polynitroaromatic explosives at lower temperature with molecular structure [18], heat of explosion [19], detonation characteristics 18, 20], and thermal stability [21]. The Arrhenius parameters are related to critical temperature of thermal explosion by Frank-Kamenetski model [22]. Thus it is important to assess the kinetic parameters for thermolysis of HEMs. Traditionally, model-fitting [23, 24] approaches are used for evaluating global activation energy and frequency factor. However the thermolysis of energetic materials is highly complex and is an intricate interplay of various physico-chemical processes. Individual steps may have different activation energies and the contribution of individual steps to the overall process may be a function of both temperature as well as extent of conversion. Such changes are not detected by traditional model fitting methods [25, 26]. Recently, the use of isocoversional methods are suggested as a solution for this problem [27, 28]. Although, there are many isoconversional methods [29-31], Vyazovkin [32] has developed a model free isoconversional method, which was successfully used to describe the thermolysis kinetics of energetic compounds such as RDX[33], HMX [34], 5-nitro-2,4-dihydro-3H-1,2,4-triazole-3-one (NTO) [35], ammonium perchlorate (AP) [36], ammonium nitrate (AN) [37] ammonium dinitramide (ADN) [25, 38] etc. The same method has been applied to evaluate kinetic parameters for composite energetic materials (CEMs) such as composite solid propellants [8, 39]. Therefore, this method has been used for determining the dependencies of activation energy on extent of conversion for the PBX samples studied. Kinetic parameters from conventional model fitting approaches have also been obtained for the isothermal decomposition of PBXs and pure energetic compounds. The merits and demerits of both modes of kinetic analyses in describing the actual mechanism of thermolysis also discussed here.

8.2. Thermolysis of HMX and Its PBXs Containing Estane

Samples of HMX, Estane and two PBX formulations containing HMX and Estane in the ratios 95:5 (HXE 9505) and 90:10 (HXE 9010) were analysed using various thermoanalytical techniques. Thermal curves from DSC analysis are given in figure 1. Pheneomenological data from the DSC analysis and TG analysis are summarized in table 3.

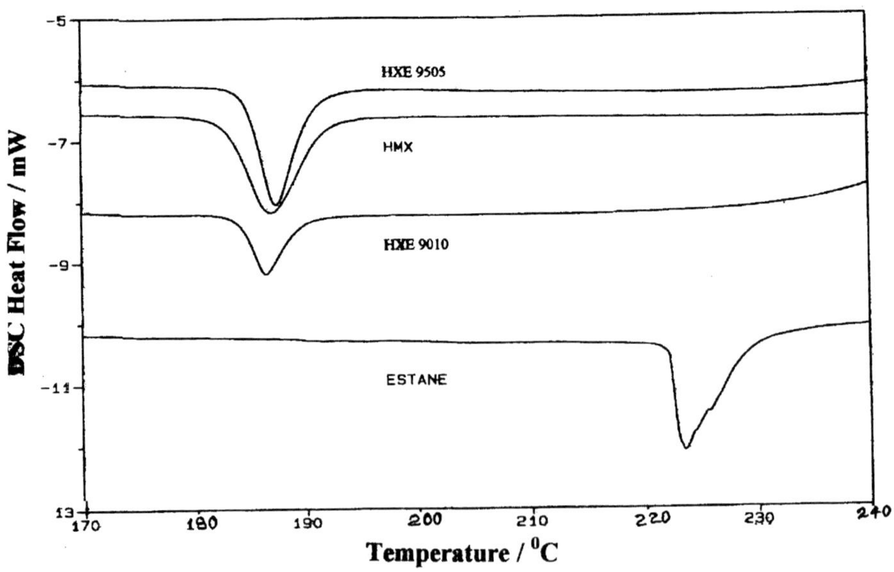

Figure 1. (a).

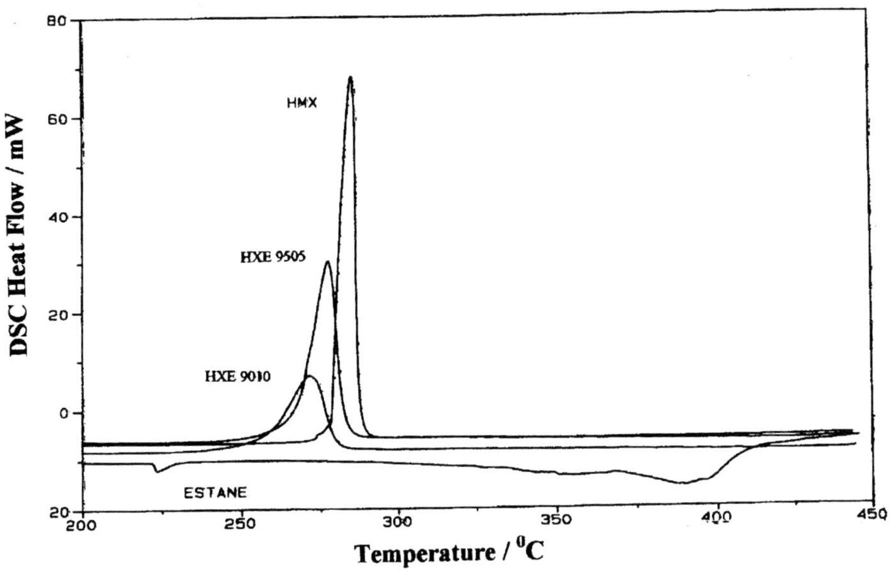

Figure 2 (b).

Figure 1. DSC thermal curves of HMX, HXE 9505, HXE 9010 and Estane under inert (N_2) atmosphere.

- in the temperature range 170 – 240 ^{0}C.
- in the temperature range 200 – 450 ^{0}C.

Table 3. TG-DTG and DSC Phenomenological Data of HMX, Estane and Their PBXs under Inert Atmosphere

Sample Name	TG-DTG		DSC Peak Temperature / ^{0}C		ΔH / J g^{-1}	
	T_i / ^{0}C	% Decomposition	Endo	Exo	Endo	Exo
HMX	278	99.97	187	286	31.6	1211
Estane	302	41.30	224	--	6.1	
	378	52.66	355	--		
			390	--	319.8	--
HXE 9505	263	97.00	187	278	28.8	1386
HXE 9010	254	94.78	186	272	25.2	1647

T_i = Onset temperature.

The results of TG-DTG and DSC studies clearly show that thermal stability of HMX is at stake when Estane is used as a binder to form corresponding PBX. Separate TG and DTA studies under reactive atmosphere also confirmed this apprehension beyond doubt. The extent of lowering in decomposition temperature for HMX increases as the percentage of binder increases. It is evident from figure 1b that the exothermic decomposition process in the case of HXE 9010 is almost complete, even before the onset temperature for the thermolysis of pure HMX and it takes place below the melting point of HMX. It is also notable that the enthalpy change (ΔH) (table 3) for the exothermic decomposition of HMX and its PBXs follows the order HXE 9010 > HXE 9505 > HMX. This may be due to the change in reaction pathways. The gasification of the binder also takes place along with that of HMX, probably due to the high heat release, and was evident from the TG-DTG data. The higher percentage consumption of binder in HXE 9010 may be due to the higher heat release.

Under the same conditions, thermal decomposition of Estane is overall endothermic in nature (figure 1b). Thermal decomposition of the Estane is reported [40] to be taking place in multi steps. In the first step, depolycondensation of the urethane linkage takes place, regenerating the diol and isocyanate. The endotherm at around 223 ^{0}C, corresponds to this process. However, there was no significant mass loss during this step in TG. Volatilization of the monomers takes place in the next step along with the formation of further byproducts through oxidation-reduction reactions. The volatilization of the products competes with the oxidation-reduction reactions. Under an inert atmosphere, it was noted that the former process is the predominant one.

Such lowering of thermal stability of energetic materials in their PBXs was observed during some earlier studies as well. Ger et al. [41] reported that when energetic binders such as polyethylene glycol (PEG) and glycidyl azide polymer (GAP) were mixed with RDX and HMX, reduction in decomposition temperature of the pure constituents took place. It was explained that the exothermic decomposition of the binder, that takes place at a lower temperature than that of the energetic filler, supplies enough heat for the early decomposition of the filler and thus thermal stability of the mixture is lowered [41]. Thermolysis of Estane is not exothermic overall, under an inert atmosphere. Thus the above said explanation does not suit for PBX samples with Estane as binder, at least in an inert atmosphere. One possible explanation may be that free radicals, that might have formed during the thermolysis of Estane, interact with HMX and thus cause its early decomposition. In such a case, the mechanism of thermolysis may change, which may be assessed by kinetic analysis. Actually, DSC thermal curves (figures 1a and 1b) show that thermal reactions in Estane start even before the onset temperature of TG. Thus these reactions that do not lead to gasification may be having some role in the early decomposition of HMX.

8.2.1. Kinetic Analysis of TG Data

Solid-state kinetics may be studied by thermal analysis. Although there is considerable skepticism against solid-state reaction kinetics by thermal analysis, if presented cautiously will give some useful indications of the mechanistic pathways. Model fitting methods are traditionally used for obtaining global activation energies of solid-state thermal decomposition reactions. Activation energy (E) for thermolysis of the HMX and its PBXs were obtained from model fitting method and the value of E for HMX obtained (≈ 200 kJ mol^{-1}) is a good estimate and is very close to the accepted value[50] of activation energy (≈ 217 kJ mol^{-1}) for the thermolysis of HMX. The values of E obtained for HXE 9505 is approximately 235 kJ mol^{-1} and is higher than that for HMX, whereas E for HXE 9010 is about 145 kJ mol^{-1}, which is lower. However the change in values of E is accompanied with a change in frequency factor. A plot of ln(A/min^{-1}) versus E was linear and thus, the change in activation energy is accurately compensated by the change in frequency factor, showing 'kinetic compensation effect'. Hence, it is very difficult to make any useful interpretation from the kinetic data obtained from model fitting methods.

Many of the disadvantages of the traditional model fitting methods can be overcome by model free isoconversional methods. These methods yield the effective activation energy as a function of extent of conversion and thus assist in detecting multi-step complex reactions. We have adopted the isoconversional method suggested by Vyazovkin [32], since the same method was used in the thermolysis of HMX [34, 44] and its composite propellant with HTPB [8]. Lofy and Wight [42] have studied the kinetics of thermolysis of HMX, using non-isothermal TG, isothermal TG, non-isothermal DSC and isothermal DSC and concluded that the global activation energy for thermal decomposition (146 ± 14 kJ mol^{-1}) is independent of extent of conversion. Our kinetic computations for the thermolysis of HMX from the isothermal TG data agree well with the above said observation. However, the average activation energy value from our study (215 kJ mol^{-1}) differs from their value. This difference might have arose from the fact that in the above said study, the thermolysis of HMX was studied below its melting point, whereas in our study the temperature range chosen was in the region of melting point of HMX. Thus the results also suggest a difference in mechanism of thermolysis of HMX below its melting point and in/above its melting point.

The results of isoconversional kinetic computations on HXE 9505 and HXE 9010 reveal that the activation energy for thermolysis is independent of extent of conversion, for these samples also. The average value of activation energy for HXE 9505 (208.4 kJ mol^{-1}) is almost same as that for pure HMX. However, the average value of E for HXE 9010 (153.2 kJ mol^{-1}) is considerably lower than that for HMX. But the decrease in activation energy cannot be attributed to any catalytic decomposition of HMX in HXE 9010. The difference in activation energy is only due to the difference in the range of temperature studied. In fact the temperature range used for studying the thermolysis of HXE 9010 was lower than the melting point of HMX. Interestingly the value of E for HXE 9010 matches well with the value reported by Lofy and Wight [41] for the thermolysis of HMX from isothermal TG studies, below its melting point. Thus the kinetic parameters from isoconversional computations do not suggest any difference in mechanism of thermolysis of HMX in its PBXs with Estane. The role of binder seems to be reducing the decomposition temperature of HMX, without affecting the energy barrier.

Activation energy of approximately 250 kJ mol^{-1} is obtained for the thermal decomposition of Estane, from model fitting methods along with a frequency factor of 42 min^{-1}. However the isoconversional method shows that the activation energy does not remain constant throughout during the thermolysis of Estane. An average value of E, 181.7 kJ mol^{-1} is observed up to around 50% conversion, which gradually changes into 244 kJ mol^{-1} afterwards. In fact DTG shows a change in the value of $d\alpha/dT$ at around 40% conversion and thus isoconversional method clearly reflects the complexity and changes in the mechanism during the thermolysis of Estane.

8.3. Thermolysis of RDX and Its PBXs Containing Estane and Viton A

Thermolysis of RDX, Estane, Viton A and their PBX formulations containing RDX and Estane in the ratios 95:5 (RXE 9505), 90:10 (RXE 9010) and a RDX-Viton A PBX in the ratio 90:10 (RXV 9010) were studied. Thermolysis of RDX involves reactions in many phases and hence the use of open and confined pans for analyses using TG and DSC, allows to vary the contributions of condensed and gas phase processes [33, 43]. The results (figure 2 and table 4) of thermal analyses at various conditions (open, partially and completely confined pans and different atmosphere) and using different techniques did not indicate any considerable reduction in the thermal stability of RDX in its PBXs. But, non-isothermal TGA under static air atmosphere showed that thermolysis of PBXs completes earlier as compared to RDX. Isothermal TGA under static air atmosphere showed that thermolysis of RDX is slower at higher temperatures where as that of PBXs occurs quite faster.

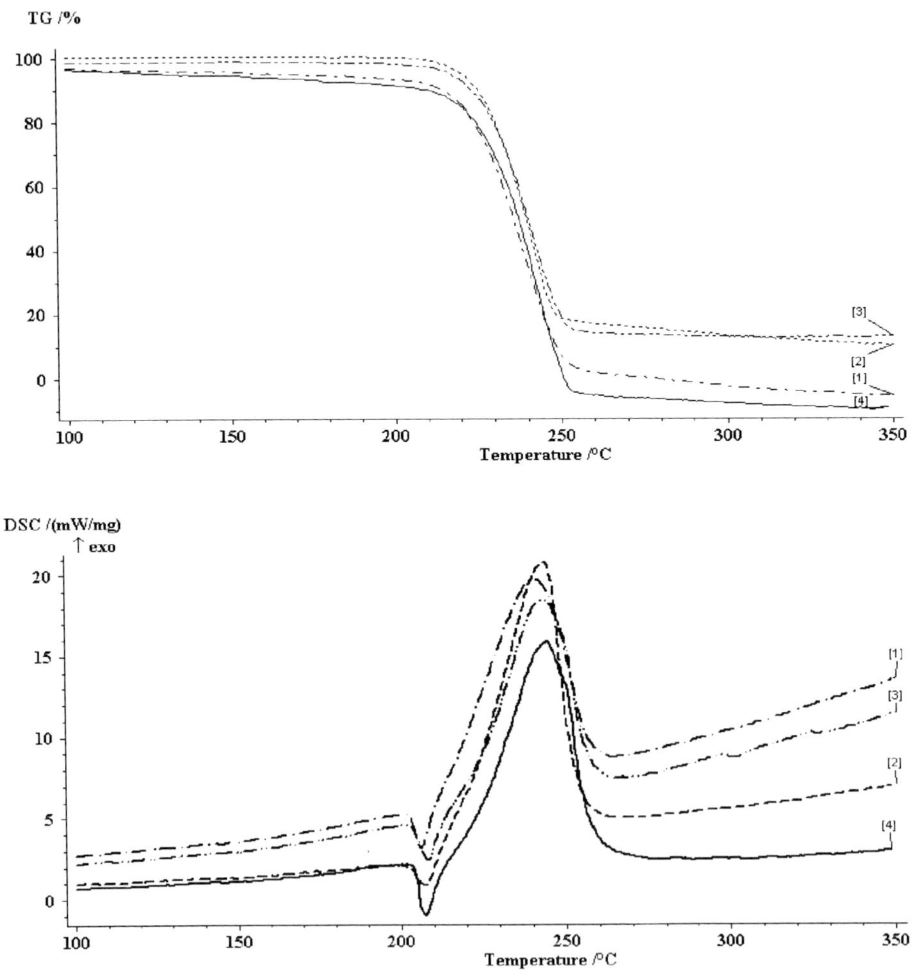

Figure 2. Simulteneous TG-DSC thermograms of RDX [4], RXE 9505 [1], RXE 9010 [2] and RXV 9010 [3] under inert atmosphere, in pierced pan.

Table 4. TG-DTG and DSC Phenomenological Data of RDX and its PBXs Under Inert Atmosphere in Open Pans

Sample Name		TG-DTG				DSC Peak Temperature / ^0C		ΔH / J g^{-1}	
		T_i / ^0C	T_s / ^0C	T_f / ^0C	% Decomposition	Endo	Exo	Endo (J g^{-1} of RDX)	Exo
RDX		216	232	240	99.7	206	236, 249*	140.5(140.5)	543
RXE 9505		214	224	235	96.7	206	239	110.5(116.3)	1216
RXE 9010		212	226	234	92.4	206	240	94.3(104.8)	1412
RXV 9010	I	219	232	239	87.1	206	240	94.0(104.4)	1179
	II	467	481	489	11.8	--	--	--	--
Viton A		462	477	485	97.8	--	--	--	--

* Shoulder peak, T_i – Onset temperature, T_s – inflection temperature, T_f – endset temperature.

RDX may initially sublimate and/or melt before eventual decomposition during heating. From figure 2, it can be seen that pure RDX sample and RDX in its PBXs shows first an endotherm and then the exotherm corresponding to decomposition. Total heat absorbed by RDX during endothermic process is higher than its PBXs, as per DSC that was run in open pans. Total heat absorbed per gram of RDX in its PBXs was recalculated, on the basis of theoretical constitution of the PBXs and given in table 4. These values also show that heat absorbed per gram of RDX is lower in the PBXs. Two possible reasons are there for this. Firstly, the contribution of the endothermic processes such as sublimation and evaporation in RDX may be high. On the contrary the heat release due to early condensed phase decomposition reactions in PBXs may cause the lowering of endothermicity for them. However, normalized peak area for the endothermic process was similar in RDX and the PBXs according to the DSC runs in partially confined pans. Endothermicity is stronger in open pans than the confined pans, irrespective of the samples.

Value of ΔH for the exothermic process for the samples as per DSC in open pans follow the order, RXE 9010 > RXE 9505 > RXV 9010 > RDX. Same order has been followed in DSC run in pierced pans. Thus, it may be seen that coating with a binder enhances ΔH for exothermic decomposition of RDX. Recently G.T. Long et al. [33] have studied the competitive vaporization and decomposition of liquid RDX. They have shown that confining an RDX sample in a closed pan prevents sublimation of solid RDX and thus size of the normalized exotherm increases as compared to that observed in open pan experiments. Thus, it seems that the polymeric binder might be forming a superficial layer over molten RDX, that prevents evaporation and thus the thermolysis proceeds in the liquid phase. This 'covering effect' is more prevalent in Estane based PBXs than the Viton A formulation. It was observed in our previous study (Section A) that depolycondensation of urethane linkage in Estane takes place, regenerating the diol and the isocyanate at ~223 ^{0}C. Thus liquid oligomers can provide better cover than Viton A, which is a thermally stable polymer.

8.3.1. Kinetic Analysis

For RDX, RXE 9505, RXE 9010 and RXV 9010, average values of ~ 157 kJ mol^{-1}, ~220 kJ mol^{-1}, ~189 kJ mol^{-1} and ~201 kJ mol^{-1} were obtained as activation energies respectively from model fitting approach. Kishore [43] has reported that an activation energy of 171±8 kJ mol^{-1} may be considered as the authentic value, for thermolysis of RDX. Our estimate of ~157 kJ mol^{-1} is near to this value. But most of the studies in melt phase of RDX reported [13, 44-47] a value of ~200 kJ mol^{-1}. Brill et al. [48] have compiled the values of activation energy reported in literature and found that all these E and lnA values usually compensate for one another. They have argued that all the E, ln(A) pairs, which lie in the regression line of their kinetic compensation plot are valid and correct for the specific characteristics of the sample and measurement used. We have plotted all the values of E for RDX and its PBXs, against corresponding ln(A) values. All the values fael in a straight line and obeyed the following equation

$$\ln(A) = a + b\,E \qquad (1)$$

Thus the kinetic parameters obtained from model fitting approach suffer from kinetic compensation effect. Further, as per the arguments of Brill et al. [48], the variation of E for the PBXs from RDX is only due to the change in sample characteristics, since our experimental conditions are the same. So it is not clear whether there is any change in mechanism for the thermolysis of RDX in its PBXs, from the kinetic parameters obtained using conventional model fitting methods. Moreover, the mechanism of thermolysis of RDX is not so simple to be defined by a single value of kinetic parameters [34].

The conventional model fitting approach fails to reveal the complexities during thermolysis. The model free isoconversional method has been recently applied [34] to study the kinetics of thermolysis of liquid RDX using non-isothermal TG and DSC techniques. Hence it was thought appropriate to use the same method for kinetic analysis of isothermal TG data of RDX and its PBXs. Activation energy for RDX changes from an initial value of 200 kJ mol^{-1} up to $\alpha = 0.25$ to ~ 150 kJ mol^{-1} at higher values of α. Long et al. [33] have explained such a variation in activation energy. The liquid state thermal decomposition of RDX is occurring through three major steps: vaporization, liquid phase decomposition and gas phase decomposition, which are having activation energies of ~100 kJ mol^{-1}, ~200 kJ mol^{-1} and ~140 kJ mol^{-1} respectively. Thus initial stages of isothermal decomposition of RDX are dominated by liquid phase decomposition. But, contribution from the other two competing channels increases as the reaction progresses and hence the reduction in activation energy.

E_α for the PBX samples is independent of α and has a value of ~ 200 kJ mol^{-1} for all the three of them. This shows that liquid phase thermolysis is the dominant process throughout their thermolysis. The role of binder is to prevent evaporation of liquid RDX and thus the thermolysis proceeds in the melt phase itself. This is also supported by higher exothermic ΔH value for the PBXs.

8.4. Thermolysis of Keto-RDX and Its PBXs Containing Thermally Stable Polymers

2-oxo-1,3,5trinitro-1,3,5-triazacyclohexane (K-6) is a cyclic dinitrourea and is the keto-derivative of 1,3,5-trinitro-1,3,5-triazacyclohexane (RDX). K-6 is an interesting high explosive as it is having high density ($\rho = 1.93$ g cm^{-3}) [49] and high heat of formation (-10±1.0 kcal mol^{-1}) [50]. The compound can be prepared from cheap starting materials and is having 4% more energy than 1,3,5,7-tetranitro-1,3,5,7-tetraazacyclooctane (HMX) at early volume expansions [50]. It also finds application as a fuel component in inflatable vehicle occupant protection device with reduced smoke [51]. However, the compound is more sensitive to impact than RDX [52], HMX [50] and is thermally less stable than RDX [52]. The sensitivity of energetic compounds to hazardous stimuli may be reduced by thoroughly coating them with thermoplastic polymers to make PBXs[1]. Thus, Mitchell et al. [50] have coated K-6 with 5% Viton A to form a PBX named RX-41-AB and small scale safety tests revealed that slight reduction in sensitivity to drop hammer impact and electric spark was achieved by coating.

Samples of K-6, its PBXs with Viton A in the weight (%) ratios 95:05 (K-6V 9505), 90:10 (K-6V 9010) and with Kel F in the ratios 95:05 (K-6K 9505) and 90:10 (K-6K 9010) were analysed for their thermal stability using various thermoanalytical techniques. The TG-DTA thermograms are shown in figure 3 and the corresponding phenomenological data are summarized in table 5. Thermolysis of K-6 occurs in a single step with ~90% mass loss. No further mass loss was observed for K-6 up to 500 ^{0}C. It is also evident from table 5 that thermolysis in PBXs occurs in multi steps. The first step in the thermolysis of PBXs is invariably the decomposition of K-6. Interestingly, the onset temperature (T_i), inflection temperature (T_s) and endset temperature (T_f) for the first step in the PBXs that are summarized in table 5, are having lower values than that for pure K-6. The extent of lowering increases as the percentage of binder increases in the PBXs. Kel F [53] and Viton A [54] are fluoro polymers and hence their thermolysis occurs at high temperatures. Thus mass losses above 400 ^{0}C for the PBXs are due to the thermolysis of binders in them.

Thermolysis of K-6 was observed in a single exothermic peak in DTA at ~202 ^{0}C. The exothermic peaks corresponding to thermolysis of K-6 in the PBXs (table 5) were observed at lower temperatures than that for pure K-6. A second exothermic peak was observed above 480 ^{0}C for PBXs with Viton A. However for the PBXs with Kel F, there were no observable processes in DTA, after the first exothermic peak. Integral and normalized peak area of the samples for the DTA thermal curves are also reported in table 5. For the PBXs, the integral area of the peaks were normalized according to the theoretical mass (%) of K-6 and the binder for their respective thermolysis processes.

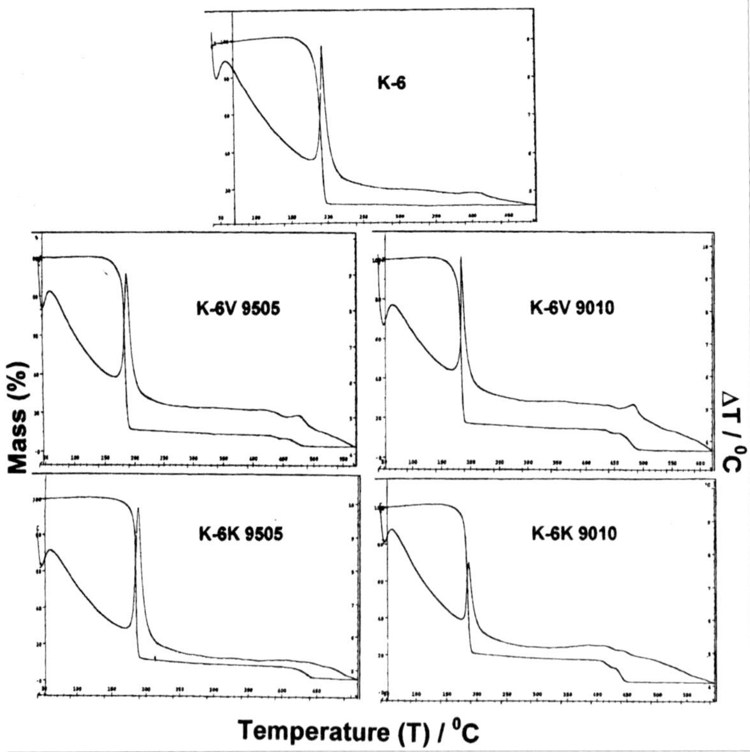

Figure 3. TG-DTA thermal curves for K-6 and its PBXs recorded under inert atmosphere at a heating rate of 10 ^{0}C min^{-1}.

The results of our thermal analyses on K-6 was similar to that are reported in literature [50, 52]. Mitchel et al. [50] have reported an exotherm at 204.9 ^{0}C for K-6 as per DSC. Our result (202 ^{0}C) is close to this value. Ostmark et al. [52] have reported the occurrence of a strong exotherm having maximum at 195 ^{0}C according to DSC. This lower value is evidently due to the lower heating rate in their study, i.e. 5 ^{0}C min^{-1}. They [52] made an ocular inspection, and reported that K-6 decomposes without melting. Mitchel et al [50] have reported that crystalline K-6 melts at 184-185 ^{0}C, without mentioning the experimental method and there was no mention of any corresponding endothermic peak in their DSC study also. The ocular inspection of Ostmark et al. [52] revealed that a gaseous bubble has interacted with the solid and formed a liquid phase. As the density of the compound is very high, it may be possible that K-6 contains strong H-bonding and hence the compound decomposes before melting. In order to assess, the occurrence of any competing reaction channels, other than solid-state decomposition, DTA analyses of K-6 were done under various conditions. No endothermic process was observed even under vacuum atmosphere (0.5 atm). Thus, we could not observe sublimation or melting and it is reasonable to conclude that thermolysis of K-6 occurs in the solid state or in the condensed phase. However, when K-6 was confined by using a sealed DTA cup, the exothermic peak was observed at slightly higher temperature.

Table 5. TG-DTA Data on K-6 and the PBXs Under Inert Atmosphere

Sample Name	TG				DTA	
	T_i / ^{0}C	T_s / ^{0}C	T_f / ^{0}C	% Mass Loss	Exothermic Peak Temperature/^{0}C	Integral Normalized Peak Area / ^{0}C mg^{-1}
K-6	191	199	203	89.6	202	183.2
K-6V 9505	186	193	197	87.9	195	229.2
	434	440	449	1.5	--	--
	464	477	486	3.7	483	450.0
K-6V 9010	185	193	195	83.2	193	214.5
	438	443	450	2.0	--	--
	470	483	492	7.8	488	397.6
K-6K 9505	187	194	197	87.8	196	280.8
	426	441	448	5.7	--	--
K-6K 9010	186	193	196	79.6	194	133.6
	413	418	426	3.6	--	--
	439	445	451	6.8	--	--

T_i – Onset temperature, T_s – inflection temperature, T_f – endset temperature.

It is interesting to note that thermally stable polymeric binders lower thermal stability of K-6. Such a lowering was observed during DSC analysis of RX-41-AB by Mitchel et al. [50]. However sensitivity of RX-41-AB was lower as compared to pure K-6, even when there was a slight decrease in thermal stability. Since, binders in the present samples are highly stable, inert fluoropolymers, the lowering in thermal stability cannot be attributed to any chemical interaction between the binder and K-6 or its decomposition products. There are probable physical reasons for the observed lowering of decomposition temperature for the PBXs, such as thermal conductivity of the samples.

The integral normalized peak areas of DTA exotherms are in the following order:

K-6K 9010 < K-6 < K-6V 9010 < K-6V 9505 < K-6K 9505

Higher heat release for the PBXs (except K-6K 9010) than K-6, show that the reaction pathways may be different. Lesser heat release for K-6K 9010 is associated with lesser mass loss (%) in the first step. Thus the actual mass (%) of K-6 may be lower in K-6K 9010 than 90 %.

Average values of 136.9 ± 7.6 kJ mol^{-1}, 135.3 ± 3.9 kJ mol^{-1}, 137.4 ± 3 kJ mol^{-1}, 143.8 ± 0.9 kJ mol^{-1} and 130.3 ± 2.9 kJ mol^{-1} were obtained as activation energies for the isothermal decomposition of K-6, K-6V 9505, K-6V 9010, K-6K 9505 and K-6K 9010 respectively in the temperature range of 170-180 ^{0}C. As per these values, the mechanism of thermal decomposition is same for K-6 and the PBXs. A plot of all values of E against corresponding ln(A) show that all these values fall almost in a straight line obeying the following equation

$$\ln(A) = 0.274\,E - 5.36 \qquad (2)$$

Ostmark et al. [52] reported 280 kJ mol^{-1} as E with a frequency factor 7.0 x 10^{29} s^{-1} in the temperature interval 175-200 ^{0}C for thermolysis of K-6, using ASTM method E 698-79 from DSC studies. This pair of E and A also obey eq 2, indicating that our kinetic parameters are consistent with values reported by Ostmark et al. using non-isothermal DSC and the difference in values are due to kinetic compensation effect.

The isoconversional analysis shows that thermal decomposition reactions of the samples are not simple as reflected by model fitting methods. K-6 has E_α values strongly dependent on values of α. E_α decreases from an initial value of ~245 kJ mol^{-1} to ~150 kJ mol^{-1} at α =1. However, the variation of E_α with α is not so drastic for the PBXs. All the four PBX samples have slightly different values of E_α, but the nature of variation in E_α with respect to α is essentially the same. The values of E_α for K-6 and the PBXs at $\alpha > 0.8$ have approximately same values. Hence it is reasonable to conclude that the final stages of thermolysis in all the samples follow the same mechanism.

Since there is an induction period and the thermolysis is taking place in the solid-state, it may be possible that initial stages (induction period) are controlled by nucleation. The acceleration period may be controlled by nuclei growth. The process of nucleation is extremely sensitive to the prehistory of a solid. Thus pure K-6 and K-6 in the PBXs may have different prehistory and hence the difference in activation energy for nucleation. In the molecular level there are many reaction pathways possible for K-6, as does RDX, which is a closely related compound in structure. Ostmark et al. [52] studied the decomposition of K-6 and RDX, using mass spectrometry (MS), laser induced mass spectrometry (LI/MS), chemiluminescence and DSC. They have concluded that concerted symmetric fission of C-N bonds is less important for K-6 than for RDX. They suggested NO_2 elimination followed by a break down of the ring as the probable mechanism of decomposition of K-6. For the production of NO from K-6 by chemiluminescence in the temperature range 90-120 ^{0}C, activation energy of 140 kJ mol^{-1} and a frequency factor of 9 x 10^9 s^{-1} was obtained. In our study E_α was ~150 kJ mol^{-1} above $\alpha > 0.8$ for K-6 and the PBXs and hence final stages of thermolysis may be limited by breaking of N-NO_2 bond. Ostmark et al. [52] have calculated bond breaking energies for N-NO_2 and concerted triple fission (C-N) of RDX and K-6, using modified neglect of diatomic overlap (MNDO) using PM3 parameter set at the unrestricted Hartree-Fock (UHF) level of theory. They have reported ΔH_{bond1} = 26.4 kcal mol^{-1} (110.5 kJ mol^{-1}) and ΔH_{bond2} = 21.5 kcal mol^{-1} (90 kJ mol^{-1}) as bond scission energies for the two types of (non-equivalent) N-N bonds in K-6. The bond scission energy for concerted triple fission was calculated to be ΔH_{bond} = 67.1 kcal mol^{-1} (280 kJ mol^{-1}). But the values calculated for RDX, i.e. ΔH_{bond} for N-N = 22 kcal mol^{-1} (92.1 kJ mol^{-1}) and ΔH_{bond} for concerted triple fission = 68.2 kcal mol^{-1} (285.4 kJ mol^{-1}) were quite different from the experimental values (~ 200 kJ mol^{-1}) for the former [55, 56] and 150 kJ mol^{-1} for the latter [57, 58] processes respectively. If the calculated values of Ostmark et al. [52] are correct, then the initial stage of K-6 decomposition may be controlled by concerted triple fission and as the reaction progresses, the contribution from N-NO_2 bond breaking increases. In the PBXs the contribution from concerted triple fission is considerably less and the other process dominates prevalently at the final stages.

REFERENCES

[1] H. Singh in *Propellants and Explosives Technology*, S. Krishnan, S. R. Chakravarthy, S. K. Athithan, Eds.: Allied Publishers Ltd.: Chennai, (1998), p 245.

[2] C. B. Storm, J. R. Stine, J. F. Kramer *Sensitivity Relationship in Energetic Materials in Chemistry and Physics of Energetic Materials*, Kluwer Academic Publishers, Dordrecht, (1990), pp.605-639.

[3] F. Chaves, J. C. Gois, P. Simoes, *Proc. 27th Int. Pyrotech. Semin.* (2000), p. 865.

[4] M. Lal, S. R. Nayak, R. Narang, S. N. Singh *J. Armt. Studies*, , *XXVII(2)* (1992) 142.

[5] M. S. Campbell, D. Garcia, D. Idar, *Thermochim. Acta*, 89 (2000) *357-358*,.

[6] A. S. Tompa, R. F. Boswell, *Thermochim. Acta,* 169 (2000) *357-358*,.

[7] M. J. Metzger, S. M. Nicolich, D. A. Geiss, R. L. Hatch, K. Y. Lee, *Proc. Int. Annu. Conf. ICT*, (1999), p 4/1-4.

[8] Paterson, G. D.; Wight, C. A. Proc. 28th NATAS Annu. Conf. Therm. Anal. Appl., 2000, p 121.

[9] Y. Oyumi, K. Inokami, K. Yamazaki, K. Matsumoto, *Propellants, Explos., Pyrotech. 18 (*1993) 62.

[10] M. W. Beckstead, *Pure and Appl. Chem. 65* (1993) 297.

[11] J. J. Lee, C. J. Tang, T. A. Litzinger, *Combust. Flame 117(4)* (1999) 795.

[12] L. D. Luca, F. Cozzi, G. Germinian, I. Ley, A. A. Zenin, *Combust. Flame 118* (1999) 248.

[13] Y. Oyumi, Proepllants, Explos., Pyrotech. 13 (1988) 42.

[14] S. N. Singh, S. S. Samudra, J. S. Gharia, Proc. 2nd Int. Autumn Semin. Propellants, Explos. Pyrotech., (1997) 112.

[15] J. S. Lee, C. K. Hsu, Proc. 28th NATAS Annu. Conf. Therm. Anal. Appl., (2000) p. 383.

[16] M. A. Cook, G. S. Horsley, W. S. Patridge, Ursenbach, W. O. *J. Chem. Phys.* 24 (1956) 60.
[17] M. A. Cook, E. B. Mayfield, W. S. Patridge, *J. Chem. Phys.* 59 (1975) 675.
[18] S. Zeman, *Thermochim. Acta* 49 (1981) 219.
[19] S. Zeman, M. Dimur, S. Truchilk, *Thermochim. Acta* 78 (1984) 181.
[20] S. Zeman, *Thermochim. Acta* 41 (1980) 199.
[21] S. Zeman, *Thermochim. Acta* 31 (1979) 269.
[22] D. A. Frank-Kamenetskii, *Acta Physiochem. USSR* 10 (1939) 365.
[23] M. E. Brown, D. Dollimore, A. K.Galway, *Reactions in Solid State, Comprehensive Chemical Kinetics*; Elsevier: Amsterdam, 22 (1980).
[24] J. Sestak, Thermophysical Properties of Solids, in Comprehensive Analytical Chemistry; Elsevier: Amsterdam, 12D (1984).
[25] S. Vyazovkin, C. A. Wight, *J. Phys. Chem.A* 101 (1997) 8279.
[26] S. Vyazovkin, C. A. Wight, *Int. Rev. Phys. Chem.* 17 (1998) 407.
[27] M. E. Brown, M. Maciejewski, S. Vyazovkin, R. Nomen, J. Sempere, A. Burnham, J. Opfermann, R. Strey, H. L. Anderson, A. Kemler, R.Keuleers, J. Janssens, H. O. Desseyn, C. R. Li, T. B. Tang, B. Roduit, J. Malek, T. Mitsuashi, *Thermochim. Acta* 355 (2000) 125.
[28] S. Vyazovkin, C. A. Wight, *Annu. Rev. Phys. Chem.* 53 (1997) 340-341.
[29] J. Friedman, *J. Polym. Sci.* 6C (1963) 183.
[30] T. Ozawa, *Bull. Chem. Soc. Japan 38* (1965) 1881.
[31] J. H. Flynn, L. A. Wall, *J. Res. Nat. Bur. Standards 70A* (1966) 487.
[32] S. Vyazovkin, *Int. J. Chem. Kinet.* 28 (1996) 95.
[33] G. T. Long, S. Vyazovkin, B. A. Breams, C. A. Wight, *J. Phys. Chem. B* 104 (2000) 2570.
[34] S. Vyazovkin, C. A. Wight, *Thermochim. Acta* 53 (1999) *340-341*,.
[35] G. T. Long, B. A. Breams, C. A. Wight, *Proc. 28th NATAS Annu. Conf. Therm. Anal. Appl.*, (2000) p 109.
[36] S. Vyazovkin, C. A. Wight, *Chem. Mater.* 11 (1999) 3386.
[37] S. Vyazovkin, J. S. Clawson, C. A. Wight, *Chem. Mater.* 13 (2001) 960.
[38] S. Vyazovkin, C. A. Wight, *J. Phys. Chem. A* 104 (1997) 7217.
[39] T. Sell, S. Vyazovkin, C. A. Wight, *Combust. Flame 119* (1999) 174.
[40] B. Suryanarayana, R. Graybush, *Ind. Chim. Belge.*, (1967) 32.
[41] M. D. Ger, W. H. Hwu, C. C. Huang, *Thermochim. Acta, 224* (1993) 127.
[42] ofy and C. A. Wight, "Thermal Decomposition of HMX Below its Melting Point" Personal Communication.
[43] K. Kishore, *Propellants Explos. 2* (1977) 78.
[44] R. N. Rogers, L. C. Smith, *Thermochim. Acta* 1 (1970) 1.
[45] R. N. Rogers, *Thermochim Acta 9* (1974) 444.
[46] F. C. Rauch, A. J. Fanelli, *J. Phys. Chem.* 73 (1969) 1604.
[47] M. A. Schreoder, Proc. 17th Jannaf Combust. Meet., CPIA Publications, 329: Sept Vol. II (1980) p 493.
[48] T. B. Brill, P. E. Gongwer, G. K. Williams, *J. Phys. Chem. 98* (1994) 12242.
[49] R. Gilardi, J. L. Flippen-Anderson, C. George, *Acta Crystallogr. Sect. C.* 4 (1990) 706.
[50] A. R. Mitchell, P. F. Pagoria, C. L. Coon, E. S. Jessop, J. F. Poco, C. M. Traver, R. D. Breithaupt, G. F. Moody, *Propellants, Explos., Pyrotech. 19 (*1994) 232.

[51] H. R. Blomouist, *U.S. Patent Appl. No.* 203113, (1998).
[52] H. Ostmark, H. Bergman, K.Ekvall, A. Langlet, *Thermochim. Acta 260* (1995) 201.
[53] S. Madorsky, S.Stauss, *J. Res. Natl. Std.* 55 (1955) 223.
[54] Section 8.3 of this chapter.
[55] C. F. Melius, J. S. Binkley, *Proc. 21st Int. Symp. Combust.*, (1986), p 1953.
[56] C. F. Melius, *in Chemistry and Physics of Energetic Materials*; S. N. Bulusu, Ed.; Kluwer: London, (1990).
[57] X. Zhao, E. J. Hinsta, Y. T. Lee, *J. Chem. Phys. Chem. 88* (1988) 801.
[58] T. D. Sewell, D. L. Thompson, *J. Phys. Chem. 95* (1991) 6228.

INDEX

A

absorption, 16, 101, 127
academic, 155
accelerator, 72
acceptor, 32, 42, 43, 49
acceptors, 38, 42
accidental, vii, 4, 6, 139, 153
accounting, 67
accuracy, 19
acetate, 12, 129
acetic acid, 84
acetone, 54, 134
acetylene, 2
acid, vii, 8, 9, 28, 31, 32, 33, 39, 41, 53, 65, 71, 72, 73, 74, 77, 79, 84, 85, 111, 121, 124, 125
acidic, 80, 121, 126, 127, 139
acoustic, 140
activation, 25, 26, 30, 36, 44, 50, 57, 65, 67, 80, 82, 99, 100, 106, 107, 111, 114, 115, 116, 119, 136, 137, 143, 147, 155, 158, 159, 162, 163, 166
activation energy, 25, 26, 30, 36, 44, 50, 57, 65, 67, 80, 82, 99, 100, 106, 107, 111, 114, 115, 116, 119, 136, 137, 143, 147, 155, 158, 159, 162, 163, 166
active centers, 29
additives, 8, 10, 23, 30, 38, 39, 40, 41, 42, 43, 44, 45, 46, 49, 50, 84, 85, 90, 130, 148, 155
adipate, 8, 44
adsorption, 19
aerosol, 11, 65
Ag, 49, 84, 85, 122, 142
ageing, 9, 136, 140
agent, 24, 54
agents, 4, 8, 71
air, 15, 17, 80, 81, 89, 95, 96, 97, 102, 103, 104, 105, 107, 109, 111, 112, 114, 116, 130, 136, 139, 141, 142, 143, 160
aircraft, 6, 61
aliphatic amines, 121
aliphatic compounds, 73
alkali, 84, 125, 141
alkaline, 77, 125, 127, 141, 144
alkylation, 89
alkylation reactions, 89
alloys, 11, 16, 17
alternative, 12
aluminium, 7, 11, 129
amine, 2, 8, 10, 46, 55, 64, 65, 93, 116, 119, 121, 122, 126, 139, 143, 147, 148
amines, 54, 56, 73, 139
amino, 6, 8, 64, 72, 73, 74, 80, 126
amino groups, 72, 74
ammonia, 10, 17, 26, 28, 31, 32, 33, 34, 41, 63, 73, 74, 101, 107, 126
ammonium, 4, 9, 11, 17, 23, 24, 25, 26, 27, 30, 32, 34, 43, 46, 51, 53, 55, 56, 57, 59, 61, 62, 64, 65, 69, 79, 85, 90, 129, 139, 146, 156
ammonium salts, 23, 26, 59, 85
analytical techniques, 12, 13
aniline, 10, 74, 107
anion, 23, 25, 26, 29, 32, 34, 40, 42, 56, 67, 90, 100, 111
anions, 41, 79, 89, 95, 100, 111
anisotropy, 30
anthracene, 11
application, 2, 10, 61, 80, 89, 126, 130, 139, 140, 146, 155, 163
aqueous solution, 64, 85, 90, 125, 126, 127
aqueous solutions, 125
ARC, 120, 136
argument, 40
aromatic, 2, 71, 73, 74, 78, 79, 85, 121
aromatic compounds, 73

arrest, 23, 29
Arrhenius parameters, 136, 155
assessment, 20, 140
ASTM, 166
atmosphere, 14, 15, 16, 17, 25, 27, 62, 81, 82, 97, 101, 104, 105, 106, 107, 109, 110, 111, 112, 115, 116, 136, 141, 142, 143, 157, 158, 160, 164, 165
atoms, 24, 84, 121, 126, 128
attention, 49, 80, 84, 120, 140
autocatalysis, 138

B

barium, 41, 47, 124
barrier, 138, 139, 159
basicity, 57, 67, 79
batteries, 11
behaviours, 137
benchmark, 3
benefits, 73
benzene, 73, 77, 120
billets, 4, 154
biuret, 112
black, 8, 12
boiling, 10, 16
bomb, 11, 147
bonding, 67, 72, 74, 85, 89, 124
bonds, 84, 101, 130, 138, 166
Boron, 11
branching, 54
breakdown, 11, 131, 132
brittle polymer, 128
bubble, 165
burn, 3, 11, 12, 72, 140
burning, 6, 11, 12, 61, 84, 85, 90, 130, 140
burns, 61

C

Ca^{++}, 42
cadmium, 125
calcium, 41, 47
caliber, 129
calibration, 19
capacity, 13
carbamide, 42, 43
carbide, 8
carbon, 2, 8, 65, 72, 99, 126, 131, 132
carbon atoms, 126
carbon dioxide, 131
carbon monoxide, 65
carbonates, 48, 85, 90, 119, 141
carboxylic, 74
carrier, 14
cast, 7
catalyst, 38, 40, 41, 49, 71
catalysts, 8, 17, 39, 40, 44, 49, 89, 90, 140
catalytic, 17, 38, 39, 40, 41, 89, 130, 136, 140, 143, 159
catalytic activity, 38, 39, 40, 41, 130, 140
catalytic effect, 143
catalytic properties, 89
cation, 23, 25, 29, 30, 32, 34, 35, 40, 41, 42, 49, 56, 89, 128
cations, 89
cavities, 33
cell, 24, 121
cellulose, 8
CH4, 73, 77
channels, 137, 138, 163, 165
charcoal, 7
chemical, vii, 1, 2, 3, 6, 7, 11, 12, 13, 16, 17, 18, 19, 20, 26, 29, 30, 34, 35, 37, 90, 111, 126, 129, 130, 131, 137, 155, 165
chemical approach, 6
chemical composition, 30
chemical energy, 1, 20, 90
chemical interaction, 34, 166
chemical reactions, 30, 34
chemicals, 2, 10
chemiluminescence, 120, 131, 136, 137, 166
chemistry, vii, 61, 71, 72, 77, 78, 131, 155
chloride, 4, 24, 101, 154
chlorine, 2, 24, 30, 61
chromatographic technique, 119
chromium, 40, 44, 45, 46
cis, 111
classes, vii, 2, 7, 8, 38
classification, 3, 9, 12
classified, 2, 3, 7, 10, 11, 38, 121, 128, 139, 147, 148
cleavage, 65, 74, 84, 131, 138, 141
clusters, 148
C-N, 65, 68, 73, 77, 78, 84, 120, 126, 127, 128, 130, 131, 132, 136, 138, 148, 166
CO2, 15, 17, 28, 47, 55, 57, 59, 64, 65, 69, 72, 73, 77, 80, 85, 90, 119, 120, 125, 131, 132, 134, 136, 139, 148
cobalt, 39, 49, 107, 111, 112, 113, 114
colors, 90, 107
combustion, 7, 9, 28, 44, 61, 62, 130, 139, 140, 147
combustion chamber, 9
combustion stability, 130, 140
commercial, 12, 64
compatibility, 10, 77
compensation, 18, 119, 136, 153, 159, 162, 163, 166

compilation, 121, 122, 126
complexity, 107, 116, 159
components, 9, 14, 90
composite, 9, 23, 84, 85, 90, 130, 139, 140, 153, 155, 156, 159
composition, 11, 12, 14, 25, 72, 78, 125, 129, 132
compositions, 1, 6, 7, 11, 12, 38, 74, 84, 90, 130
compounds, vii, 1, 2, 3, 7, 14, 30, 47, 55, 56, 61, 64, 67, 71, 72, 73, 74, 78, 79, 80, 83, 84, 85, 89, 90, 121, 139, 140, 141, 147, 148, 153, 156, 163
concentration, 30, 35, 43, 50, 89, 130, 131
condensation, 80
conduction, 32
conductivity, 18, 23, 24, 25, 26, 50, 89, 166
configuration, 112, 123
conjugation, 6, 124
consensus, 119, 120, 148
construction, 4
consumption, 158
control, 57, 67
controlled, 12, 15, 166
convective, 12
conversion, 18, 80, 82, 99, 116, 156, 159
cooling, 8
coordination, 89, 100, 107, 112
copolymer, 4, 154, 155
copolymers, 129
copper, 8, 44, 45, 46, 100, 101, 103, 104, 105, 106, 109, 111, 112, 141
copper oxide, 44, 105, 141
correlation, 136
correlation coefficient, 136
correlations, 155
corrosive, 10
coupling, 20
covering, 162
crack, 4, 153
cracking, 10
CRC, 69, 117
creep, 10
critical temperature, 119, 128, 146, 148, 155
crystal, 4, 23, 24, 25, 28, 29, 30, 34, 35, 37, 38, 43, 50, 72, 74, 77, 101, 112, 121, 124, 127, 146, 153
crystal growth, 24
crystal lattice, 24, 25, 72, 74
crystal structure, 101, 112, 127
crystal structures, 127
crystalline, 18, 19, 23, 54, 63, 72, 100, 107, 112, 121, 127, 165
crystallisation, 124, 141
crystallites, 50
crystallization, 16, 18, 24, 26, 38, 43, 125
crystals, 4, 7, 24, 26, 28, 35, 37, 41, 50, 64, 127, 153, 154
curing, 18
cutters, 5, 155
cyanamide, 72
cyanide, 141
cycling, 26

D

DBP, 8
decomposition, 1, 4, 6, 14, 16, 17, 23, 26, 27, 28, 29, 30, 32, 33, 34, 35, 36, 38, 39, 40, 41, 42, 43, 44, 46, 47, 48, 49, 50, 51, 53, 54, 55, 56, 57, 59, 63, 64, 65, 67, 68, 69, 72, 73, 74, 77, 78, 80, 84, 85, 89, 90, 95, 98, 99, 100, 101, 103, 107, 108, 109, 111, 112, 116, 119, 120, 121, 124, 125, 130, 131, 132, 133, 134, 135, 136, 137, 138, 139, 141, 142, 143, 148, 153, 155, 156, 157, 158, 159, 162, 163, 164, 165, 166
decomposition reactions, 53, 84, 85
decomposition temperature, 32, 50, 57, 64, 67, 98, 111, 124, 157, 158, 159, 166
defects, 24, 32
definition, 12
deformation, 34, 37
degradation, 15, 80, 131
degree, 12, 30
dehydration, 14, 16, 85, 101, 119, 141, 142, 143, 146, 148
dehydrogenation, 89
delays, 11, 12
demand, 5, 155
denitrification, 141
density, 4, 6, 10, 12, 24, 50, 74, 75, 78, 79, 82, 83, 119, 124, 154, 163, 165
Department of Energy, 4, 150, 154
dependant, 6, 63
derivative thermogravimetry, 15
derivatives, 15, 72, 74, 75, 77, 80, 82
desorption, 14, 133
detachment, 25
detection, 12, 15, 84, 132
detonation, 3, 4, 6, 7, 12, 73, 78, 129, 155
diamines, 67, 78
diazonium salts, 77
dielectric, 13, 32
differential scanning, 12, 19, 120
differential scanning calorimeter, 19
Differential Scanning Calorimetry (DSC), 12, 13, 17, 18, 19, 20, 27, 62, 65, 74, 77, 80, 82, 93, 94, 95, 98, 100, 104, 105, 107, 108, 110, 115, 120, 128,

129, 131, 136, 137, 140, 141, 142, 143, 147, 156, 157, 158, 159, 160, 161, 162, 163, 164, 165, 166
Differential Thermal Analysis (DTA), 12, 13, 15, 16, 17, 18, 20, 36, 50, 55, 57, 65, 67, 74, 79, 80, 81, 83, 93, 94, 95, 96, 97, 98, 100, 101, 102, 103, 104, 105, 106, 107, 108, 109, 110, 111, 112, 113, 114, 115, 120, 128, 131, 136, 140, 141, 142, 157, 164, 165, 166
differentiation, 15, 20
diffraction, 24, 101, 124
diffusivity, 18
dimer, 84, 139
dimeric, 112
dislocation, 24, 30, 37
dislocations, 24, 29, 30, 33, 34, 35, 37, 50, 51
displacement, 35, 37, 74
dissociation, 34, 56, 65, 111, 138
distilled water, 90
DMA, 13
DNI, 45, 79
donor, 26, 38, 42, 43
DOP, 5, 8, 154
dopant, 42
doped, 39, 41, 42, 49
doping, 26, 37, 39, 40, 42, 49
DOS, 155
DPO, 80
DRS, 12
dry, 23, 53, 54
DSC method, 147
DTA curve, 17, 36, 50, 81, 97, 104, 109, 113, 115
duration, 11
dyes, 11
dysprosium, 125

E

earth, 7, 10, 84, 122, 125, 127, 141, 144
EDA, 95, 96, 102, 108, 112
electrical, 12, 89, 100, 147
electrical conductivity, 12, 89, 100
electron, 24, 28, 29, 32, 35, 40, 74, 78, 85, 124, 131
electron microscopy, 29
electron paramagnetic resonance, 28, 35
electronic, 15, 78
electrons, 32
electrostatic, 12, 124
endothermic, 16, 17, 54, 80, 95, 98, 101, 103, 104, 113, 115, 116, 141, 142, 158, 162, 165
energetic materials, vii, 1, 2, 6, 7, 20, 78, 119, 124, 131, 153, 155, 158
energetic parameters, 126
energy, 2, 4, 6, 8, 11, 16, 18, 20, 30, 46, 56, 67, 73, 74, 78, 79, 80, 83, 84, 99, 100, 108, 111, 119, 126, 129, 132, 136, 138, 139, 140, 143, 145, 146, 147, 148, 153, 156, 158, 159, 162, 163, 167
engines, 61
entropy, 139
environment, 5, 10, 14, 139
environmental, 12, 89
EPR, 134, 135
equilibrium, 29, 31
equipment, 15
ESR, 24, 120
ESR spectroscopy, 120
esters, 2
estimating, 129
ETA, 13
etching, 24
ethane, 8
ethanol, 24, 54, 65
ethylene, 80, 129, 143
ethylenediamine, 89, 95, 101, 103, 108, 111, 126, 129
evaporation, 32, 54, 77, 127, 162, 163
evidence, 28, 42, 84, 137
evolution, 1, 14, 17, 39, 68, 73, 85, 125, 132, 136, 139
excitation, 34, 35
exothermic, 16, 17, 46, 73, 77, 80, 90, 95, 98, 101, 105, 107, 108, 112, 113, 114, 116, 119, 128, 137, 141, 142, 143, 148, 157, 158, 162, 163, 164, 165
exothermic effects, 16
exothermic peaks, 164
expansions, 163
experimental condition, 28, 63, 69, 119, 130, 163
experts, vii
explosive, 1, 2, 3, 4, 5, 6, 7, 9, 12, 56, 61, 63, 64, 65, 72, 73, 74, 77, 78, 79, 80, 82, 83, 84, 90, 101, 103, 119, 120, 121, 128, 129, 130, 139, 153, 154, 155, 163
explosives, vii, 1, 2, 3, 4, 6, 7, 11, 12, 20, 53, 61, 63, 64, 65, 71, 72, 73, 74, 78, 79, 80, 84, 85, 90, 120, 129, 136, 139, 147, 153, 154, 155
exponential, 143

F

fabrication, 16, 17
fertilizers, 61
fillers, 155
fission, 65, 166
flame, 1, 8, 12, 62, 90, 139, 145, 147
flight, 5, 135, 155
flow, 10

fluoride, 4, 154
fluorides, 38
fluorine, 74
fluoropolymers, 165
formaldehyde, 65, 72, 80, 83, 84
formamide, 74
Fourier, 132
fragmentation, 28, 56, 131
free radical, 57, 132, 133, 158
free radicals, 133, 158
friction, 1, 3, 6, 12, 55, 56, 57, 63, 64, 67, 73, 77, 80, 90, 145, 147, 153
FTIR, 63, 65, 108, 120, 137, 141
FTIR spectroscopy, 120, 141
fuel, 4, 6, 7, 9, 10, 11, 12, 44, 53, 56, 78, 163
fusion, 13, 62

G

gamma radiation, 36
gamma rays, 36
gas, 1, 12, 13, 14, 25, 27, 28, 32, 33, 34, 40, 41, 61, 63, 65, 73, 84, 124, 128, 134, 139, 140, 148, 160, 163
gas chromatograph, 27
gas phase, 25, 28, 32, 33, 34, 40, 41, 63, 65, 84, 124, 139, 140, 160, 163
gases, 1, 89, 131, 132
gasification, 27, 158
gels, 10
generation, 37, 132
geometrical parameters, 121
Ger, 158, 168
glass, 18, 40
glass transition, 18
glycol, 8, 158
grains, 10
granules, 129
gravimetric analysis, 12
groups, 1, 2, 3, 20, 56, 67, 71, 72, 73, 74, 77, 83, 90, 119, 127, 128, 131
growth, 29, 30, 31, 32, 33, 42, 166
growth rate, 30, 31
guidance, 90

H

H_2, 15, 55, 73, 74, 77
halogen, 1, 74
handling, 1
H-bonding, 127, 165
health, 12
health problems, 12
heat, 1, 3, 11, 12, 14, 16, 18, 19, 20, 39, 48, 49, 57, 73, 74, 77, 78, 84, 90, 101, 119, 131, 139, 140, 145, 147, 155, 158, 162, 163, 166
heat capacity, 18, 19, 20
heat release, 90, 158, 162, 166
heat transfer, 12, 140
heating, 1, 14, 15, 16, 17, 19, 27, 28, 36, 39, 46, 48, 74, 77, 83, 90, 95, 100, 103, 107, 111, 119, 132, 135, 137, 141, 142, 148, 162, 164, 165
heating rate, 14, 15, 16, 17, 19, 27, 36, 46, 95, 101, 107, 111, 132, 142, 164, 165
heavy metal, 12, 85
helium, 111
heterocycles, 80
heterogeneous, 8, 30
hexafluoropropylene, 4, 154
high pressure, 1, 4, 10, 153
high resolution, 28
high temperature, 1, 4, 27, 34, 36, 90, 100, 132, 138, 154, 164
high-speed, 128
homogeneity, 12
homogeneous, 8
homolog, 147
horse, 23
HPLC, 120, 136, 137
humidity, 15
hybrid, 10
hydrate, 54
hydrates, 84, 119, 125, 141, 146, 148
hydrazine, 10, 12, 54, 77, 80
hydro, 73
hydrocarbon, 65, 71
hydrocarbons, 73
hydrogen, 10, 32, 67, 71, 72, 74, 85, 124, 126, 132, 134, 138, 139
hydrogen atoms, 134
hydroxide, 85, 125
hydroxides, 85
hydroxyl, 2, 34, 83, 130, 140
hypothesis, 132

I

ICT, 129, 150, 167
identification, 134
IDP, 8
illumination, 3, 11
impact energy, 72
impurities, 19
IMS, 131
incompatibility, 12

indication, 30, 98
indices, 39
induction, 30, 31, 35, 55, 166
induction period, 30, 31, 35, 55, 166
industry, 11
inert, 10, 16, 18, 19, 27, 62, 116, 157, 158, 160, 164, 165
infrared, 90, 108
inhibition, 39, 47
inhibitors, 38, 47
initiation, 4, 5, 6, 17, 77, 128, 129, 130, 138, 153, 155
innovation, 18
inorganic, 1, 2, 11, 14, 53, 61
insight, 138
inspection, 165
instability, 8, 64, 141
instruments, 18
integrity, 129
intensity, 28
interaction, 32, 33, 34, 38, 44, 148
intermolecular, 124, 134
interpretation, 17, 19, 159
interval, 15, 101, 166
ionic, 127
ionization, 34, 131, 133, 135, 138
ions, 12, 24, 32, 34, 35, 37, 41, 42, 49, 50, 89
IR, 91, 92, 120, 126, 127, 131, 132, 136, 137, 140
IR spectra, 127
IR spectroscopy, 120, 131, 132
iron, 49
irradiation, 23, 34, 35, 36, 37, 38, 43
irradiations, 50
isomers, 73, 77, 141
isothermal, 12, 35, 65, 67, 74, 77, 85, 99, 100, 102, 103, 105, 106, 111, 116, 136, 137, 143, 147, 156, 159, 160, 163, 166
isotope, 132

K

Keynes, 151
kinetic constants, 137
kinetic curves, 43
kinetic parameters, 48, 112, 119, 120, 130, 136, 137, 142, 143, 155, 159, 163, 166
kinetic studies, 15
kinetics, 14, 51, 67, 73, 74, 89, 99, 111, 114, 136, 137, 155, 158, 159, 163

L

language, vii
laser, 28, 130, 131, 132, 133, 166
lattice, 25, 26, 29, 30, 32, 33, 37, 38, 41, 42, 49, 57, 145, 146
lattices, 34
lead, 8, 12, 32, 34, 35, 37, 73, 85, 141, 158
learning, vii
ligand, 89, 90, 98, 101, 104, 105, 108, 109, 111, 115, 116
ligands, 89, 93, 95, 100, 101, 107, 116
linear, 5, 14, 33, 136, 155, 158
linear defects, 33
linkage, 158, 162
liquid fuels, 10
liquid oxygen, 7
liquid phase, 30, 162, 163, 165
liquor, 12, 90
literature, vii, 42, 55, 89, 90, 95, 100, 121, 122, 124, 128, 139, 146, 155, 162, 164
lithium, 39, 124, 141
location, 19
losses, 101, 164
low temperatures, 29
low-temperature, 124
LTD, 23, 27, 30, 38, 39, 50, 54, 55
lubricants, 8
luminescence, 35

M

magnesium, 7
magnetic, 89, 112
manganese, 114, 115
manifolds, 14
mass loss, 15, 27, 44, 80, 93, 94, 98, 99, 101, 104, 112, 158, 164, 166
mass spectrometry, 55, 65, 131, 166
mass transfer, 140
matrix, 4, 9, 153
measurement, 14, 15, 19, 57, 100, 162
mechanical, 4, 13, 23, 24, 30, 38, 50, 129, 153
mechanical properties, 129
melamine, 72
melt, 63, 65, 129, 162, 163
melting, 18, 19, 26, 57, 63, 64, 65, 66, 72, 74, 75, 77, 80, 112, 124, 128, 139, 158, 159, 165
melting temperature, 63
melts, 63, 65, 69, 74, 165
metal hydroxides, 125

metal oxide, 11, 12, 23, 38, 46, 49, 89, 98, 108, 119, 140, 141, 148
metal oxides, 23, 38, 89, 119, 140, 141, 148
metal salt, 84, 85
metal salts, 71, 85, 119, 121, 124, 125, 126, 127, 139, 140, 141, 143, 144, 146, 147, 148
metals, 85, 121, 126, 143
methanol, 65, 90
methyl group, 56, 59, 64, 65, 72, 77
methyl groups, 56
methylation, 83
microscopy, 131
migration, 26, 42, 132, 139
military, 3, 9, 11, 84, 139
mining, 3, 4
missiles, 7
mixing, 12, 38, 53, 55, 79, 125, 126
MNDO, 167
model fitting, 80, 89, 99, 116, 153, 156, 158, 159, 162, 163, 166
models, 30
moieties, 101
moisture, 12, 14, 73
molar ratio, 40
mole, 8
molecular dynamics, 121
molecular structure, 73, 155
molecular weight, 6, 72, 73
molecules, 1, 6, 33, 74, 78, 89, 90, 107, 108, 110, 112, 119, 121, 134, 138, 143, 148, 153
monograph, 19
monomers, 158
motors, 11, 61
movement, 29

N

nanocrystalline, 89
National Bureau of Standards, 69
Nd, 84, 122, 127, 142, 146
neglect, 167
neutrons, 36
NHC, 75
Ni, 11, 38, 49, 85, 91, 92, 93, 94, 101, 107, 108, 109, 110, 111, 114, 115, 122, 142
nickel, 39, 49, 107, 110, 111
nickel oxide, 39, 110
NiO, 38, 39, 40, 41, 107, 110
nitrate, 2, 4, 7, 8, 11, 12, 26, 44, 61, 62, 63, 64, 65, 66, 67, 69, 72, 79, 89, 90, 91, 93, 95, 97, 98, 100, 101, 103, 104, 106, 107, 108, 109, 110, 111, 112, 114, 115, 116, 139, 147, 156
nitrates, 61, 64, 65, 67, 68, 69, 71, 90, 111, 125
nitric acid, vii, 53, 61, 63, 64, 65, 71, 90, 121, 125
nitric oxide, 65, 136
nitrite, 133
nitrogen, 28, 44, 57, 61, 63, 65, 71, 105, 106, 107, 110, 112, 124, 128, 132, 134, 147
nitrogen dioxide, 65, 71
nitrogen oxides, 28, 44
nitrous oxide, 63, 131
NMR, 24, 126, 127
N-N, 75, 84, 167
NO, 1, 17, 27, 28, 34, 53, 54, 55, 56, 59, 64, 65, 68, 72, 73, 74, 76, 77, 78, 80, 84, 85, 89, 119, 120, 122, 127, 131, 132, 133, 134, 136, 137, 138, 148, 166
noise, 17, 100
normal, 3, 126
normal conditions, 3
nucleation, 166
nuclei, 29, 30, 31, 33, 37, 42, 43, 49, 54, 166
nucleus, 1, 29, 30, 31, 32, 34, 68

O

observations, 101, 126
oil, 4, 7
oligomers, 162
optical, 11, 29
organ, 2
organic, 1, 2, 7, 8, 11, 13, 43, 54, 56, 63, 107
organic compounds, 54
organic peroxides, 2
organometallic, 2
orthorhombic, 23, 24, 26, 27, 50, 55, 59, 72, 127
oxalate, 44
oxidation, 13, 16, 17, 30, 34, 39, 40, 41, 56, 57, 63, 65, 67, 68, 73, 77, 78, 80, 89, 98, 101, 103, 107, 112, 132, 158
oxidative, 18, 85, 116, 132
oxide, 8, 9, 39, 40, 41, 46, 47, 49, 74, 82, 89, 112, 115, 125, 141
oxides, 38, 39, 48, 49, 85, 89, 140
oxygen, 2, 9, 17, 24, 30, 53, 61, 62, 65, 83, 89, 128, 135
oxygenation, 134
ozone, 9, 136
ozonides, 2

P

paper, 150
paraffins, 72
paramagnetic, 28

parameter, 167
particles, 4, 153
pathways, 31, 58, 59, 68, 69, 95, 100, 105, 119, 120, 121, 130, 131, 133, 134, 135, 138, 140, 148, 155, 158, 166
Pb, 48, 84, 85, 122, 141, 146
PCT, 151
PDP, 57
perchlorate, 2, 7, 9, 11, 12, 23, 25, 26, 27, 30, 32, 33, 39, 43, 46, 49, 51, 53, 54, 55, 56, 57, 59, 61, 90, 101, 129, 139, 156
performance, vii, 4, 6, 9, 11, 20, 61, 78, 79, 80, 85, 119, 124, 128, 129, 136, 148, 155
permeation, 12
peroxide, 2, 7, 9
petroleum, 90
pH, 54
phase decomposition, 137, 162, 163
phase diagram, 18, 129
phase transformation, 18
phase transitions, 112
phenol, 77
phosphorus, 7
photochemical, 130, 134
photon, 133, 135
physical properties, 61
physico-chemical properties, 41
planar, 138
plastic, vii, 4, 7, 84, 129, 139, 153, 154
platinum, 16, 17
play, 13, 42, 138
PM3, 167
pollutant, 89
pollutants, 89
polyamides, 141
polybutadiene, 130, 140
polyethylene, 158
polymer, 4, 7, 18, 129, 132, 153, 154, 158, 162
polymer matrix, 4
polymeric materials, 11
polymeric products, 72, 85
polymerization, 81
polymers, 4, 7, 11, 18, 129, 153, 163, 164
polyurethane, 155
pore, 37
pores, 28, 29, 30, 34
porosity, 12
porous, 12, 29, 30
porous materials, 12
potassium, 7, 11, 12, 124, 141
potential energy, 1, 139
powder, 7, 11, 12, 101, 124, 129
powders, 3, 7, 11
power, 3, 6, 7, 18, 73, 77
precipitation, 146
prediction, 136
preparation, 4, 12, 55, 79, 80, 89, 121, 124, 125, 126, 139, 154
pressure, 5, 6, 12, 30, 31, 33, 34, 62, 63, 80, 130, 132, 137, 155
primary products, 28, 50
probability, 4, 32
procedures, 6, 84, 90
producers, 11, 12
production, 6, 32, 63, 67, 139, 148, 166
productivity, 19
progressive, 103
propagation, 1, 11, 73, 129
property, iv, 13, 17, 18
proportionality, 18
propulsion, 7, 10
protection, 163
protons, 36, 42
PTA, 12
pulsed laser, 135
pulses, 4, 153
pumps, 10, 61
pyrolysis, 14, 28, 130, 132, 140, 141

Q

quantitative estimation, 18
quantum, 138
quantum mechanics, 138

R

radiation, 34, 35, 36, 37, 51, 130
radical, 32, 35, 73, 131, 134
radius, 85
Raman, 117
range, 2, 32, 39, 56, 63, 77, 80, 85, 93, 94, 98, 101, 108, 132, 136, 137, 143, 148, 157, 159, 166
rare earth, 38, 125, 127, 141
rare earths, 38
rationalisation, 143
raw material, 12
raw materials, 12
reactant, 12
reaction rate, 12, 30, 50
reaction zone, 12
reactivity, 37, 51
real time, 133
recombination, 63

reduction, 13, 40, 65, 67, 73, 77, 80, 98, 103, 153, 158, 160, 163
reflection, 153
regression, 10, 162
regression line, 162
relationship, 136
reliability, 14
research, vii, 13, 80, 119, 120, 121, 126, 128, 136, 139, 155
research and development, 13
researchers, 120
residues, 80, 90, 105, 130, 140
resilience, 10
resistance, 84, 119
resolution, 17, 19
rhodium, 17
rhombohedral, 24
rings, 99
room temperature, 48, 61, 69, 99, 134

S

safety, vii, 5, 12, 20, 130, 136, 155, 163
salt, 11, 12, 39, 65, 79, 84, 85, 119, 124, 125, 141, 143, 146, 147, 148
salts, vii, 8, 12, 44, 53, 56, 57, 61, 63, 64, 65, 66, 67, 79, 80, 84, 85, 90, 111, 119, 120, 121, 122, 124, 125, 126, 127, 129, 130, 139, 140, 141, 142, 143, 146, 147, 148
sample, 13, 14, 15, 16, 17, 18, 19, 28, 35, 36, 46, 47, 119, 124, 136, 148, 162, 163
sampling, 133
satellite, 23
sawdust, 7
scanning calorimetry, 13
Schmid, 151
school, 120
Schottky, 32
science, 20
scientific, vii
scientists, 124, 146, 147
screw dislocations, 24
seeds, 29
segregation, 10
selectivity, 29
sensitivity, 4, 6, 7, 12, 15, 17, 18, 19, 20, 55, 57, 63, 65, 67, 72, 73, 74, 75, 77, 79, 80, 83, 84, 119, 128, 129, 130, 139, 145, 147, 148, 153, 163, 165
sensors, 89
separation, 5, 19, 155
series, 14, 17, 18, 39, 65, 84, 111, 116, 133, 139
services, 139
shape, 29
shear, 10, 133
shock, 1, 3, 4, 6, 50, 54, 55, 74, 78, 80, 90, 131, 133, 153
signaling, 11
signals, 11, 28, 131
silica, 7, 16
silver, 49, 124, 141, 146
similarity, 49
sintering, 15
SiO2, 38
sites, 24, 29, 33, 37, 42, 83, 126
Sm, 84, 122, 127, 142, 146
smoke, 11, 163
sodium, 85, 124, 141, 146
software, 18
solid phase, 38, 55, 57, 65, 124
solid state, 27, 30, 65, 99, 100, 137, 165
solid-state, 137, 158, 165, 166
solutions, 125, 126
solvent, 14
species, 17, 89, 100, 103, 132, 134, 141
specific heat, 19
specific surface, 40, 41, 89
spectra, 131, 133, 134
spectroscopy, 12, 65, 120, 126
spectrum, 89, 126, 131, 139
stability, 6, 12, 14, 73, 78, 100, 111, 119, 133, 136, 138, 143, 148, 158, 165
stabilizers, 7, 8
stages, 15, 25, 30, 40, 47, 56, 65, 101, 103, 137, 163, 166
standards, 168
stimulus, 2, 5
stoichiometry, 12
storage, 5, 10, 13, 26, 38, 43
strain, 30, 79
strains, 15
strategy use, 6
strength, 4, 9, 10, 153
strontium, 41, 47
structural changes, 16
students, vii
substances, 1, 2, 12, 20, 73
substitution, 56
sulfur, 7
sulphate, 44
sulphur, 2
superposition, 28
surface area, 15, 42, 43
surface layer, 38, 50
surface reactions, 13, 40
symmetry, 30
synthesis, 41, 72, 74, 77, 78, 80, 83, 119, 129

synthetic, 77, 78
systematic, 78, 90, 153, 155
systems, 5, 10, 78, 130, 139, 155

T

tanks, 9
TDI, 8
teaching, vii
technological, 53, 74
temperature, 8, 9, 10, 11, 12, 14, 15, 16, 17, 18, 19, 20, 23, 25, 26, 27, 29, 30, 32, 33, 34, 35, 39, 45, 46, 47, 49, 53, 55, 56, 59, 62, 63, 65, 67, 98, 100, 101, 108, 114, 125, 129, 130, 132, 134, 135, 136, 137, 138, 139, 140, 142, 143, 145, 147, 155, 157, 158, 159, 161, 164, 165, 166
temperature dependence, 18, 129, 135
tetrahydrofuran, 155
tetroxide, 9
TGA, 12, 13, 14, 20, 46, 47, 48, 105, 120, 136, 137, 160
theoretical, 4, 25, 119, 120, 124, 138, 139, 147, 148, 154, 162, 164
theory, 167
thermal, vii, 6, 12, 13, 14, 17, 18, 23, 25, 26, 27, 28, 29, 30, 31, 34, 35, 36, 37, 38, 39, 40, 41, 42, 43, 44, 47, 48, 49, 50, 53, 55, 56, 57, 58, 59, 62, 63, 65, 67, 68, 69, 72, 73, 77, 78, 79, 80, 82, 83, 84, 85, 93, 95, 98, 99, 100, 101, 106, 111, 112, 114, 115, 119, 120, 121, 124, 130, 131, 132, 133, 136, 137, 138, 140, 141, 142, 143, 146, 147, 148, 153, 155, 157, 158, 159, 160, 163, 164, 165, 166
thermal activation, 120, 130
thermal analysis, 12, 13, 120, 140, 142, 158
thermal decomposition, 13, 14, 17, 23, 25, 27, 28, 29, 30, 31, 34, 35, 36, 37, 38, 39, 40, 41, 42, 43, 44, 47, 48, 49, 50, 53, 55, 57, 58, 59, 63, 65, 68, 69, 72, 73, 77, 78, 95, 101, 112, 119, 120, 130, 131, 132, 136, 137, 138, 141, 143, 155, 158, 159, 163, 166
thermal degradation, 84
thermal expansion, 62
thermal properties, 26, 55, 101
thermal stability, vii, 6, 13, 14, 26, 43, 63, 67, 73, 77, 84, 85, 95, 98, 99, 100, 106, 111, 114, 115, 124, 130, 136, 143, 147, 153, 155, 157, 158, 160, 164, 165
thermodynamic, 20, 146
thermograms, 20, 95, 96, 101, 102, 103, 106, 107, 108, 109, 110, 112, 114, 160, 164
thermogravimetric, 14, 15, 20, 36, 112
thermogravimetric analysis, 14, 20
thermogravimetry, 120
thermolysis, vii, 14, 23, 26, 27, 29, 30, 31, 35, 37, 38, 39, 49, 50, 53, 54, 56, 57, 59, 61, 63, 64, 65, 67, 69, 71, 77, 78, 80, 83, 84, 85, 89, 90, 95, 98, 99, 100, 101, 103, 105, 108, 110, 112, 114, 116, 119, 120, 130, 131, 132, 133, 134, 136, 137, 139, 140, 141, 143, 144, 145, 148, 153, 155, 157, 158, 159, 160, 162, 163, 164, 165, 166
thermo-mechanical, 12
thermoplastic, 129, 153, 163
thin film, 132, 134
thin films, 132
Ti, 8, 11, 46, 85, 157, 161, 164, 165
time, 11, 12, 13, 14, 15, 18, 25, 26, 28, 38, 39, 40, 44, 47, 50, 80, 100, 101, 135, 137
TiO_2, 39
titanium, 11
TMA, 12, 13, 65
toluene, 8, 77
total energy, 7, 140
traditional model, 156, 159
traffic, 11
trans, 111
transfer, 23, 24, 25, 26, 28, 29, 32, 33, 34, 40, 42, 46, 50, 54, 56, 57, 59, 63, 64, 65, 67, 69, 78, 120, 130, 132, 138, 143
transformation, 16, 18, 19, 20, 30, 59, 64
transformations, 1, 13, 16, 19, 62, 64
transition, 18, 19, 23, 26, 27, 29, 32, 36, 38, 49, 50, 56, 57, 61, 65, 69, 89, 90, 93, 125, 130, 138, 139, 140, 141, 147
transition metal, 38, 49, 89, 90, 93, 116, 117, 125, 140, 141, 147
transition temperature, 18, 26, 50
transitions, 19, 63, 69
transport, 18
traps, 35, 37, 42, 50
triggers, vii, 6, 130
twinning, 121

U

UHF, 167
UN, 64
uniform, 16
urea, 8, 64, 69, 72, 74, 85, 112
urethane, 158, 162
UV, 130, 133
UV radiation, 130

V

vacancies, 25, 42, 49

vacuum, 14, 28, 34, 133, 165
valence, 32
values, 30, 57, 116, 124, 126, 136, 137, 142, 143, 146, 147, 148, 158, 162, 163, 164, 166, 167
variables, 18
variation, 15, 99, 163, 166
vehicles, 5, 7, 23, 155
velocity, 6, 29, 62, 135, 155
violence, 2
violent, 55, 56, 64, 108
viscosity, 10, 130
volatile substances, 141
volatilization, 101, 158
vulnerability, 6

W

warfare, 1
water, 14, 19, 26, 29, 30, 32, 62, 63, 65, 77, 101, 110, 112, 141, 146
water vapour, 26, 30
wave number, 128
weapons, 2, 4, 6, 154
weight loss, 14, 80, 104, 105, 112, 115
wet, 4
workers, 19, 72, 85, 101, 120, 126, 136

X

X-ray, 36, 55, 101, 121, 124, 127, 130
X-ray analysis, 127
X-ray crystallography, 127
X-ray diffraction, 121, 124
X-ray photoelectron spectroscopy (XPS), 130, 131
X-rays, 36
XRD, 124, 126, 127, 131, 140

Y

yield, 17, 74, 80, 111, 134, 136, 137, 159
yttrium, 126

Z

Zen, 77
zinc, 41, 95, 98, 100
zirconium, 11
Zn, 11, 38, 48, 85, 91, 92, 93, 94, 95, 96, 97, 98, 99, 100, 105, 106, 108, 111, 114, 115, 116, 122, 142, 146
ZnO, 38, 89, 95, 98